NEW DIRECTIONS FOR YOUTH DEVELOPMENT

Theory
Practice
Research

summer | 2009

Universities in Partnership

Strategies for Education, Youth Development, and Community Renewal

Ira Harkavy
Matthew Hartley

issue
editors

Gil G. Noam
Editor-in-Chief

JOSSEY-BASS™
An Imprint of
WILEY

UNIVERSITIES IN PARTNERSHIP: STRATEGIES FOR EDUCATION, YOUTH DEVELOPMENT, AND COMMUNITY RENEWAL
Ira Harkavy, Matthew Hartley (eds.)
New Directions for Youth Development, No. 122, Summer 2009
Gil G. Noam, Editor-in-Chief
This is a peer-reviewed journal.

Microfilm copies of issues and articles are available in 16mm and 35mm, as well as microfiche in 105mm, through University Microfilms Inc., 300 North Zeeb Road, Ann Arbor, Michigan 48106-1346.

NEW DIRECTIONS FOR YOUTH DEVELOPMENT (ISSN 1533-8916, electronic ISSN 1537-5781) is part of The Jossey-Bass Psychology Series and is published quarterly by Wiley Subscription Services, Inc., A Wiley Company, at Jossey-Bass, 989 Market Street, San Francisco, California 94103-1741. POSTMASTER: Send address changes to New Directions for Youth Development, Jossey-Bass, 989 Market Street, San Francisco, California 94103-1741.

SUBSCRIPTIONS for individuals cost $85.00 for U.S./Canada/Mexico; $109.00 international. For institutions, agencies, and libraries, $228.00 U.S.; $268.00 Canada/Mexico; $302.00 international. Prices subject to change. Refer to the order form that appears at the back of most volumes of this journal.

EDITORIAL CORRESPONDENCE should be sent to the Editor-in-Chief, Dr. Gil G. Noam, McLean Hospital, Harvard Medical School, 115 Mill Street, Belmont, MA 02478.

Cover photograph by © Ally Clark/istockphoto

www.josseybass.com

Contents

Issue Editors' Notes

OVER THE PAST TWO DECADES, a democratic, engaged, civic university movement has developed across the United States. A central feature of this movement has been university-community partnerships in which higher educational institutions work with organizations and schools in their local community. Much of this work has focused on the education and development of young people.

The extent to which colleges and universities have developed serious, sustained, democratic, mutually beneficial partnerships that genuinely benefit the community and its residents has varied widely. As tends to be the case in the early days of any movement, rhetoric at times outstrips reality. Nevertheless, significant change has occurred regarding both the quantity and quality of partnerships, and intriguing models have been developed. Still, significant challenges remain. The struggle to achieve transformative democratic practice in the face of seemingly intractable obstacles resides at the heart of this volume of *New Directions for Youth Development*.

Developing powerful and effective university-community partnerships for youth development is extremely difficult to do. It requires, among other things, overcoming traditional ivory tower thinking and doing; developing creative, comprehensive approaches; and engaging in long-term democratic, collaborative work. Fortunately, a number of partnerships fit this bill and serve as helpful examples to other higher educational institutions and their communities. We chose five such university-community partnerships from across the United States to feature in this volume. Each has been developed over a number of years and has focused on making a genuine difference in the condition of young people and their schools and communities. The case studies are from the

NEW DIRECTIONS FOR YOUTH DEVELOPMENT, NO. 122, SUMMER 2009 © WILEY PERIODICALS, INC.
Published online in Wiley InterScience (www.interscience.wiley.com) • DOI: 10.1002/yd.301

1

State University of New York, Buffalo; Indiana University-Purdue University Indianapolis; University of Pennsylvania; University of Dayton; and Widener University. Academics, community and school partners, and university administrators, including a university president (Widener University), have authored or coauthored the articles.

All of the articles, including ours, focus on university-school partnerships. Given the significance of the university-school connection and the obvious relationship between education and youth development, it is logical that schooling and education would be consistent themes across case studies. Even so, we did not foresee it when we planned the volume. We contacted colleagues who were involved with comprehensive, multidimensional partnerships. Given what we knew of their work, we assumed that only the University of Pennsylvania and University of Dayton cases would extensively discuss schools. That all the article authors chose to highlight projects with public schools is a powerful indicator that university-school-community partnerships have become the primary means for developing the democratic, engaged, civic university dedicated to advancing the quality of life and learning for all Americans, particularly its children.

Ira Harkavy
Matthew Hartley
Editors

IRA HARKAVY *is associate vice president and director of the Netter Center for Community Partnerships at the University of Pennsylvania.*

MATTHEW HARTLEY *is associate professor of education at the University of Pennsylvania.*

Executive Summary

Chapter One: University-school-community partnerships for youth development and democratic renewal

Ira Harkavy, Matthew Hartley

Democratic partnerships of universities, schools, and an array of neighborhood and community organizations are the most promising means of improving the lives of our nation's young people. Over the past two decades, many colleges and universities have been experiencing a renaissance in engagement activities. Universities, once ivory towers, have increasingly come to recognize that their destinies are inextricably linked with their communities. Authentic democratic partnerships have three characteristics: they are devised to achieve democratic purposes, the collective work is advanced through inclusive and democratic processes, and the product these partnerships produce benefits all participants and results in a strengthening of the democratic practices within the community.

NEW DIRECTIONS FOR YOUTH DEVELOPMENT, NO. 122, SUMMER 2009 © WILEY PERIODICALS, INC.
Published online in Wiley InterScience (www.interscience.wiley.com) • DOI: 10.1002/yd.302

Chapter Two: The connection: Schooling, youth development, and community building—The Futures Academy case

Henry Louis Taylor Jr., Linda Greenough McGlynn

Universities, because of their vast human and fiscal resources, can play the central role in assisting in the development of school-centered community development programs that make youth development their top priority. The Futures Academy, a K–8 public school in the Fruit Belt, an inner-city neighborhood in Buffalo, New York, offers a useful model of community development in partnership with the Center for Urban Studies at the State University of New York at Buffalo. The goal of the project is to create opportunities for students to apply the knowledge and skills they learn in the classroom to the goal of working with others to make the neighborhood a better place to live. The efforts seek to realize in practice the Dewey dictum that individuals learn best when they have "a real motive behind and a real outcome ahead."

Chapter Three: George Washington Community High School: Analysis of a partnership network

Robert G. Bringle, Starla D. H. Officer, Jim Grim, Julie A. Hatcher

After five years with no public schools in their community, residents and neighborhood organizations of the Near Westside of Indianapolis advocated for the opening of George Washington Community High School (GWCHS). As a neighborhood in close proximity to the campus of Indiana University-Purdue University Indianapolis, the Near Westside and campus worked together to address this issue and improve the educational success of youth. In fall 2000, GWCHS opened as a community school and now thrives as a national model, due in part to its network of community relationships. This account analyzes the development of the school by focusing on the relationships among the university, the high school, community organizations, and the residents of the Near Westside and highlights the unique partnership between the campus and school by defining the relational qualities and describing the network created to make sustainable changes with the high school.

Chapter Four: The Agatston Urban Nutrition Initiative: Working to reverse the obesity epidemic through academically based community service

Francis E. Johnston

The Agatston Urban Nutrition Initiative (AUNI) presents a fruitful partnership between faculty and students at a premier research university and members of the surrounding community aimed at addressing the problem of childhood obesity. AUNI uses a problem-solving approach to learning by focusing course activities, including service-learning, on understanding and mitigating the obesity culture.

Chapter Five: Dayton's Neighborhood School Centers

Dick Ferguson

When Dayton Public Schools committed to return to neighborhood K–8 schools, the community organized to refocus many youth programs in schools and neighborhoods. This article describes the planning and implementation of Dayton's Neighborhood School Centers. Special emphasis is placed on the role of the University of Dayton, especially the Fitz Center for Leadership in Community. The Fitz Center plays a pivotal role in implementing this highly collaborative effort, including project leadership; community organizing; coaching of five site coordinators at neighborhood school sites; and faculty-mentored student interns to assist with programming for student success, family support, health and team sports, and extensive service-learning coordination. The Dayton Foundation, Dayton Public Schools, City of Dayton, Montgomery County, and sixteen foundation and corporate supporters are partners with the Fitz Center in a bold initiative to reconnect five Dayton public elementary schools to their neighborhoods, after more than thirty years of court-ordered busing, and create full-service, year-round opportunities for neighborhood families and youth at these new schools. More than forty programs have been started at each site, all emphasizing youth and community assets.

Chapter Six: The president's role in advancing civic engagement: The Widener-Chester partnership

James T. Harris III

Efforts by metropolitan universities to engage in meaningful and democratic partnerships with community organizations require much time, effort, and considerable resources from the university and its various constituents. Widener University is located in a distressed urban environment. This study, presented from the perspective of the university's president, highlights the challenges associated with engaging in such work and provides insight into possible future directions for advancing an institution-wide civic engagement agenda. It outlines in detail the initiatives created between Widener and the Chester, Pennsylvania, school district over six years and explains how after many failures, the university came to the conclusion that its best chance for success would be to develop a separately chartered university partnership school. The account forcefully underscores that the costs associated with civic engagement are worth the investment in spite of the number of setbacks and frustrations inherent in this type of work.

NEW DIRECTIONS FOR YOUTH DEVELOPMENT • DOI: 10.1002/yd

Democratic partnerships involving universities, schools, and an array of neighborhood and community organizations are the most promising means of improving the lives of our nation's young people and strengthening our communities.

1

University-school-community partnerships for youth development and democratic renewal

Ira Harkavy, Matthew Hartley

Democracy must begin at home, and its home is the neighborly community.
John Dewey, *The Public and Its Problems* (1927)

Democracy has been given a mission to the world, and it is of no uncertain character. . . . I wish to show that the university is the prophet of this democracy, as well as its priest and its philosopher; that in other words, the university is the Messiah of the democracy, its to-be-expected deliverer.
William Rainey Harper,
The University and Democracy (1899)

GROWING UP IN AMERICA's cities is exceedingly and unnecessarily difficult for too many children. School systems in major metropolitan

NEW DIRECTIONS FOR YOUTH DEVELOPMENT, NO. 122, SUMMER 2009 © WILEY PERIODICALS, INC.
Published online in Wiley InterScience (www.interscience.wiley.com) • DOI: 10.1002/yd.303

areas struggle to serve their needs. In our own city of Philadelphia, for example, only about 55 percent of the students entering ninth grade graduate from high school in four years. The infant mortality rate in Philadelphia rose 8 percent from 2004 to 2005, reaching 11.3 infant deaths per 1,000 live births, the highest level since 1999. In neighborhoods in West Philadelphia not far from where we work, infant mortality rates are 15 per 1,000 as compared to 5 per 1,000 in adjacent suburban counties. Yet despite these challenges, urban communities harbor rich resources that have the potential to make a profound difference in the lives of young people. As elsewhere, these communities are filled with parents who are deeply concerned about their children's welfare and anxious to act on their behalf. Furthermore, most cities contain a university, a hospital, or both. These "eds and meds" can play a critical role as anchor institutions, providing employment to many and serving as powerful collaborators in economic, educational and civic renewal efforts.[1]

Universities are well positioned to play a role in responding to the challenges facing our nation's cities. Over half of all institutions of higher learning are located within or immediately outside urban areas.[2] Universities are resource rich. In many cities, universities and hospitals are the largest private employers. According to a recent estimate, urban colleges and universities employ 3 million people, and fully two-thirds are administrative, clerical, or support staff.[3] One study concluded, "Older core cities have a significant concentration of jobs in education and health services. . . . These industries account for over 20 percent of the jobs in the case study cities, compared to 15 percent of jobs nationally."[4] As such, they serve as anchor institutions that provide significant economic stability to their local areas.[5] This status also gives them considerable leverage in encouraging and participating in systemic reform.

The partnership imperative

One strategy to marshal all of these potential resources is through university-school-community partnerships. Over the past two

NEW DIRECTIONS FOR YOUTH DEVELOPMENT • DOI: 10.1002/yd

decades, many colleges and universities have been experiencing a renaissance in engagement activities.[6] The ill-fated efforts at urban renewal in the 1960s, whose prime instrument was often a bulldozer, have evolved into rich partnerships based on mutuality and reciprocity. Universities have increasingly come to recognize that their destinies are inextricably linked with their communities. Ernest Boyer, the highly influential former president of the Carnegie Foundation for the Advancement of Teaching, called for colleges and universities to link engagement activities with core academic functions such as teaching and research.[7] He, and many others, argued that the doors of the ivory tower ought to be opened onto the wide avenues of the community—not merely because it is laudable but because it is a superior means of fulfilling the university's mission of teaching and research.

Throughout the 1990s, in answer to this call, hundreds of universities established offices or centers aimed at encouraging partnerships with the community. Hundreds of thousands of college students participate in various community-based activities. However, a significant challenge of this work has been moving beyond limited (and at times palliative) community involvement toward the establishment of deep, lasting, democratic, collaborative partnerships aimed at addressing pressing real-world problems. It is these reciprocal and comprehensive university-community partnerships, and ones aimed particularly at youth development, that we are focus on in this volume.

The activities highlighted in the articles that follow are notable because they draw in multiple constituents from the community working in genuine collaboration with the university in democratic partnerships. Furthermore, their outcomes extend far beyond the provision of services (though that is certainly an important concern). In a real sense, these partnerships aim at revitalizing communities. In our view, they exemplify democracy in action: people working together to change their communities and society for the better. Indeed, one of their defining characteristics is that they are intent on both solving community problems and building civic capacity.

We believe these partnerships represent a radical departure from business as usual. Each of the initiatives presented in this volume faced significant challenges. In part, they struggled to overcome what Benjamin Franklin derisively called the "ancient Customs and Habitudes" of institutions of higher learning: a preference for the production and preservation of esoteric knowledge (and the status it conveys) over the pragmatic pursuit of a happier, healthier, and more democratic society.[8] In this respect, these examples make a powerful case for an engaged, civic university that aims both to improve the quality of life for young people and to advance learning and research through serious, sustained, significant democratic partnerships.

Historical roots of democratic partnerships

The presence of the schooling system in each of these accounts is telling and worth underscoring. We view schooling as the key strategic subsystem of modern societies. More than any other subsystem, it exerts the greatest influence on society as a whole. These accounts also underscore the pivotal role universities can play due to their enormous intellectual, economic, social, and human capital power when they partner with communities and their schools.

The notion of universities as embedded in and intimately connected to their communities has deep historic roots. The nation's first university, the University of Pennsylvania, was founded by Benjamin Franklin in Philadelphia, one of the largest and most important cities in the American colonies. In 1749, Franklin published a pamphlet, "Relating to the Education of Youth in Pensylvania [sic]," in which he articulated a vision of an institution predicated not on classical education for the elites but to serve all students of ability in the interest of fostering an "*Inclination* join'd with an *Ability* to serve mankind, one's country, Friends and Family" [emphasis in the original].

This notion was echoed in the land grant movement in the nineteenth century, which expanded the public system of higher education and directly tied its work to the betterment of society. We also see it at the dawn of the modern research university. William Rainey Harper, the first president of the University of Chicago (1892–1906), argued that the university was the "prophet," "the Messiah of the democracy," and his vision conceived of communities, schools, and universities as inextricably linked. In 1896, the year Dewey began the Laboratory School at Chicago, Harper proclaimed his "desire to do for the Department of Pedagogy what has not been undertaken in any other institution." When criticized by a university trustee for sponsoring a journal focused on pedagogy in precollegiate schools, Harper emphatically retorted, "As a university we are interested above all else in pedagogy."[9] He argued: "Through the school system every family in this entire broad land of ours is brought into touch with the university; for from it proceeds the teachers or the teachers' teachers."[10] Harper also believed that within local contexts, universities and schools and the community had the capacity to work together for the betterment of society. So do we.

Characteristics of truly democratic partnerships

Authentic, democratic partnerships have several key characteristics in common. In October 2004, one of us (Ira Harkavy) attended the third in a series of conferences sponsored by the Kellogg Forum on Higher Education for the Public Good, held at the Johnson Foundation's Wingspread Conference Center in Racine, Wisconsin. The conference, "Higher Education Collaboratives for Community Engagement and Improvement," assigned participants to one of several working groups. The report of the faculty and researcher working group echoed many of the themes identified in this volume. Specifically, it identified democratic purpose, process, and product as crucial for successful university partnerships with schools and communities:

- *Purpose.* "It is not possible to run a course aright when the goal itself is not rightly placed," Francis Bacon wrote in 1620. A successful partnership must be known for its democratic and civic purposes. This is in keeping with the democratic mission that served as the central animating force behind the development of the American research university. An abiding democratic and civic purpose is the rightly placed goal if higher education is to truly contribute to the public good.

- *Process.* In accordance with the purpose, a successful partnership should be democratic, egalitarian, transparent, and collegial. Higher education institutions should go beyond a rhetoric of collaboration and conscientiously work with communities, rejecting the unidirectional, top-down approaches that all too often have characterized university-community interaction. The higher education institution and the community, as well as members of both communities, should treat each other as ends in themselves rather than as means to an end. The relationship itself and welfare of the various partners should be the preeminent value, not developing a specified program or completing a research project. These are the types of collaborations that tend to be significant, serious, and sustained, lead to a relationship of genuine respect and trust, and most benefit the partners and society.
- *Product.* A successful partnership strives to make a positive difference for all partners. Contributing to the well-being of people in the community (both now and in the future) through structural community improvement (for example, effective public schools, neighborhood economic development, strong community organizations) should be a central goal of a truly democratic partnership for the public good. Research, teaching, and service should also be strengthened as a result of a successful partnership. Indeed, working with the community to improve the quality of life in the community may be one of the best ways to improve the quality of life and learning within a higher education institution.[11]

For us and all of the article authors, democracy is the heart and soul of successful, significant, sustained university-school-community partnerships.

The Penn approach

Since the work at the University of Pennsylvania began over twenty-three years ago, we have devoted particular attention to developing mutually beneficial, mutually respectful democratic partnerships between Penn and schools and communities in West Philadelphia and Philadelphia. Over time, we have come to conceptualize the work of the Netter Center for Community Partnerships, the organization that Ira Harkavy directs that develops and administers Penn's school and community partnerships, to develop university-assisted community schools as an ongoing communal participatory action research project designed to contribute simultaneously to the improvement of West Philadelphia and to the university's relationship with West Philadelphia, as well as to the advancement of learning and knowledge.

As an institutional strategy, communal participatory action research differs significantly from traditional action research. Both research processes are directed toward problems in the real world, are concerned with application, and are participatory, but they differ radically in the degrees to which they are continuous, comprehensive, and beneficial to both the organization or community studied and the university. For example, traditional action research is exemplified in the efforts developed by the late William Foote Whyte, Davydd Greenwood, and their associates at Cornell University in Ithaca, New York, to advance industrial democracy in the worker cooperatives of Mondragón, Spain. Its considerable empirical and theoretical significance notwithstanding, the research at Mondragón is not at all an institutional necessity for Cornell. By contrast, the University of Pennsylvania's enlightened self-interest is directly tied to the success of its research efforts in West Philadelphia—hence, its emphasis on, and continuing support for, communal participatory action research. In short, proximity to an easily accessible site and a focus on problems that are institutionally significant to the university encourage sustained, continuous research involvement. Put another way, strategic community problem-solving research tends strongly to develop sustained, continuous research partnerships between a university and its local community.[12]

Given its fundamental democratic orientation, the Netter Center's participatory action research project has worked toward higher levels of participation by community members in problem identification and planning, as well as in implementation. This has not been easy to do. Based on decades of Penn's destructive action and inaction involving the local community, university-community conflicts take significant effort and time to reduce. The center's work with university-assisted community schools has focused on health and nutrition, the environment, conflict resolution and peer mediation, community performance and visual arts, school and community publications, technology, school-to-career programs, and reading improvement. Each of these projects almost inevitably varies in the extent to which it engages and empowers public school students, teachers, parents, and other community members in each stage of the research process. Although it has a long way to go before it actually achieves its goal, the center's overall effort has been consciously democratic and participatory, with a goal of genuinely working *with* the community, not *on* or *in* the community.

As university-assisted community schools and related projects have grown and developed and as concrete positive outcomes for schools and neighborhoods have continued to occur, community trust and participation have increased. It would be terribly misleading, however, if we left the impression that town-gown collaboration has completely—or even largely—replaced the town-gown conflicts that strongly characterized Penn-community relationships before 1985; it has not.

Since 1985, Penn's engagement with West Philadelphia schools and neighborhoods has certainly come a long way. But Penn still has a far distance to travel before it radically changes its hierarchical culture and structure and uses its enormous resources to help transform West Philadelphia into a democratic, cosmopolitan, neighborly community and a multidimensional asset for a major university. Stated directly, we do not think we have largely solved the problem of developing and implementing the practical means needed to realize Dewey's theory of participatory democracy. We are well aware that we are a long way from having done so. But we have found that working with the community to solve strategic community-identified problems is a powerful means for advancing ongoing, increasingly democratic relationships between Penn and the schools and communities in West Philadelphia.

The major component of the neo-Deweyan strategy now being developed and slowly implemented by Penn focuses on developing university-assisted community schools designed to help educate, engage, activate, and serve all members of the community in which the school is located. The strategy assumes that community schools, like colleges and universities, can function as focal points to help create healthy urban environments and that both universities and colleges function best in such environments. Somewhat more specifically, the strategy assumes that public schools, like colleges and universities, can function as environment-changing institutions and become strategic centers of broadly based partnerships that engage a wide variety of community organizations and institutions. Since public schools belong to all members of the community, they should serve all its members. (No implication is intended that public schools are the only community places where learning takes place; obviously, it also takes place in libraries, museums, private schools, and other institutions. Ideally, all the learning places in a community would collaborate.)

More than any other institution, we contend, public schools are particularly well suited to function as neighborhood hubs or centers around which local partnerships can be generated and developed. When they play that innovative role, schools function as community institutions par excellence. They then provide a decentralized, democratic, community-based response to rapidly changing community problems. In the process, they help young people learn better, at increasingly higher levels, through action-oriented, collaborative, real-world problem solving.

For public schools to function as integrating community institutions, however, local, state, and national governmental and nongovernmental agencies must be effectively coordinated to help provide the resources community schools need in order to play the greatly expanded roles we envision in American society. How to conceive that organizational revolution, let alone implement it, poses extraordinarily complex intellectual and social problems. But as Dewey forcefully argued, working to solve complex, real-world problems is the best way to advance knowledge and learning, as well as the general capacity of individuals and institutions to advance knowledge and learning.

We therefore contend that American universities should give high priority—arguably their highest priority—to solving the problems inherent in the organizational revolution we have sketched. If universities were to do so, they would demonstrate in concrete practice their self-professed theoretical ability to simultaneously advance knowledge, learning, and societal well-being. They would then satisfy the critical performance test proposed in 1994 by the president of the State University of New York at Buffalo, William R. Greiner: that "the great universities of the twenty-first century will be judged by their ability to help solve our most urgent social problems."[13]

Since 1985, to increase Penn's ability to help solve America's most urgent social problems, we have worked to develop and implement the idea of university-assisted community schools. We emphasize *university assisted* because community schools require far more resources than traditional schools do and because we have become convinced that, in relative terms, universities constitute the strategic sources of broadly based, comprehensive, sustained support for community schools.

The university-assisted community school idea we have been developing at Penn since 1985 essentially extends and updates Dewey's theory that the neighborhood school can function as the core neighborhood institution—the core institution that provides comprehensive services, galvanizes other community institutions and groups, and helps solve the problems communities confront in a rapidly changing world. Dewey recognized that if the neighborhood school were to function as a genuine community center, it would require additional human resources and support. But to our knowledge, he never identified universities as the (or even a) key source of broadly based, sustained, comprehensive support for community schools.

It is critical to emphasize, however, that the university-assisted community schools now being developed at Penn have a long way to go before they can effectively mobilize the potentially powerful, untapped resources of their communities, and thereby enable children and families to function as community problem solvers, as well as deliverers and recipients of caring, compassionate, local services. Nonetheless, the work at Penn, as well as the impressive, comprehensive university-community-school partnerships and university-assisted community schools developed at Widener University, the State University of New York at Buffalo, University of Dayton, and Indiana University-Purdue University at Indianapolis are, we are convinced, indicators of the ongoing development of the engaged, democratic, civic university that advances learning as it works with its community to realize the democratic promise of America for all Americans, particularly its children and youth.

Notes

1. Harkavy, I., & Zuckerman, H. (1999). *Eds and meds: Cities' hidden assets.* Washington, DC: Brookings Institution.

2. Hahn, A., with Coonerty, C., & Peaslee, L. (2003). *Colleges and universities as economic anchors: Profiles of promising practices.* Waltham, MA: Brandeis University, Heller Graduate School of Social Policy and Management, Institute for Sustainable Development/Center for Youth and Communities and POLICYLINK.

3. Hahn, with Coonerty & Peaslee. (2003).

4. Fox, R. K., & Trouhaft, S. (2006). *Shared prosperity, stronger regions: An agenda for rebuilding America's older core cities.* Oakland, CA: PolicyLink.

5. Harkavy & Zuckerman. (1999).

6. Hartley, M., & Hollander, E. (2005). The elusive ideal: Civic learning and higher education. In S. Fuhrman & M. Lazerson (Eds.), *Institutions of democracy: The public schools.* New York: Oxford University Press.

7. Boyer, E. (1994, March 9). Creating the new American college. *Chronicle of Higher Education.*

8. Best, J. H. (1962). *Benjamin Franklin on education.* New York: Teachers College, Columbia University.

9. R. McCaul, quoted in White, T. W. (1977). *The study of education at the University of Chicago,1892–1958.* Unpublished doctoral dissertation, University of Chicago. p. 15.

10. Harper, W. R. (1905). The university and democracy. In W. R. Harper (Ed.), *The trend in higher education* (pp. 1–343). Chicago: University of Chicago Press.

11. Harkavy, I. (2005). Higher education collaboratives for community engagement and improvement: Faculty and researchers' perspectives. In P. A. Pasque, R. E. Smerek, B. Dwyer, N. Bowman, & B. L. Mallory (Eds.), *Higher education collaboratives for community engagement and improvement* (pp. 22–23). Ann Arbor, MI: National Forum for Higher Education and the Public Good.

12. Benson, L., Harkavy, I., & Puckett, J. (2007). *Dewey's dream: Universities and democracy in an age of education reform.* Philadelphia: Temple University Press.

13. Greiner, W. R. (1994). In the total of all these acts: How can American universities address the urban agenda? *Universities and Community Schools, 4*(1–2), p. 12.

IRA HARKAVY *is associate vice president and director of the Netter Center for Community Partnerships at the University of Pennsylvania.*

MATTHEW HARTLEY *is associate professor of education at the University of Pennsylvania.*

University-assisted school-centered community development programs can produce young people who are not only better students but will also mature into caring and engaged citizens.

2

The connection: Schooling, youth development, and community building—The Futures Academy case

Henry Louis Taylor Jr.,
Linda Greenough McGlynn

THE DEVELOPMENT OF inner-city youth, especially African Americans and Latinos, is not a priority in the United States. As a society, we have chosen instead to pathologize or criminalize many social problems, strengthen the criminal justice system, and place the jailing of troubled youth as a higher-priority solution than remediation and education, which are much more likely to develop productive citizens.

This policy approach has had a devastating impact on African American and Latino students in particular. In a pioneering study of high school graduation rates, Jay P. Greene found that only 56 percent of African Americans and 54 percent of Latinos graduated from high school nationally.[1] Today the situation is so severe that one in ten high schools in the United States are now considered

NEW DIRECTIONS FOR YOUTH DEVELOPMENT, NO. 122, SUMMER 2009 © WILEY PERIODICALS, INC.
Published online in Wiley InterScience (www.interscience.wiley.com) • DOI: 10.1002/yd.304

"dropout factories," meaning that these are places where no more than 60 percent of the freshmen make it to their senior year.[2]

Blacks and Latinos are also overrepresented in prisons and jails. According to the Bureau of Justice Statistics, a branch of the U.S. Department of Justice, 10 percent of the black male population between eighteen and thirty-four years of age are in jail or prison.[3] The nationwide rate at which black youth receive life-without-parole sentences (6.6 per 10,000) is estimated at ten times greater than the rate for white youth (0.6 per 10,000). Of black males aged twenty-five to twenty-nine, 8.4 percent are sentenced inmates, compared to 2.5 percent of Latino males and 1.2 percent of white males in that age group.[4] These high incarceration and sentencing rates not only reflect the reality that the United States is the world's leading jailer; it reflects the nation's failure to invest in the development of African American and Latino youth.

Young people of color, who are at the greatest risk of dropping out of school or being jailed, or both, most often live in distressed inner-city neighborhoods. This suggests that neighborhoods matter in the development of young people and can contribute to dropping out of school, drug abuse, crime, unemployment, poverty, and a variety of other socioeconomic problems.[5]

This point is made very clearly in a 1994 *New York Times* article on the deplorable state of Philadelphia public schools. The reporter told of the heroic efforts of Rebecca Kimmelman, a newly appointed principal, to improve instruction and the overall academic environment at Meade Elementary School. Kimmelman, the reporter said, believed that ". . . teaching is not Meade's biggest problem . . ."; that distinction belongs to the distressed community in which the school exists. Kimmelman said, "You could give me $80 billion to improve the school, but it won't make much difference unless you make changes out there [in the neighborhood]. If a 6-year old's mother is a drug addict and a prostitute and she's dying of AIDS and she's all but abandoned the child, what can we do to turn that child around?"[6]

NEW DIRECTIONS FOR YOUTH DEVELOPMENT · DOI: 10.1002/yd

Neighborhoods, as this example illustrates, can increase a young person's vulnerability to a host of problems.[7] Conversely, neighborhoods that are functioning well can lower a young person's risk by creating a communal environment supportive of a healthy life and culture. Thus, depending on the character of the neighborhood place, the types of institutions located in it, the relationships that exist between institutions and residents, and the relations that exist between neighborhoods and government, neighborhoods can either protect or increase the vulnerability of young people. Thus, turning distressed inner-city neighborhoods into cosmopolitan, socially functional communities that are based on the principles of participatory democracy, reciprocity, collaboration, and social justice will certainly increase the probability that its residents, specifically black and Latino youth, will become caring, productive, and engaged citizens who will not add to the already too high incarceration and dropout statistics.

At this juncture, universities can play a critical role. Because of their vast human and fiscal resources, higher education institutions can contribute to both youth development and the transformation of inner-city neighborhoods. But first the university must forge genuine partnerships with public schools and distressed communities. Using, as an example, a case study of Futures Academy, a public school for kindergarten to eighth grade in the Fruit Belt, an inner-city neighborhood in Buffalo, New York, this article demonstrates that universities can play a leading role in remediating the problems of public schooling, youth development, and inner-city distress.

Through the development of authentic, democratically based partnerships among universities, schools, and communities, young people in distressed neighborhoods can become successful students and engaged citizens who work to improve conditions in their neighborhoods, city, and nation.[8]

This article first provides an overview of the history of university-community partnerships, sets out a review of pedagogical theories, and ends with a discussion of the pedagogical model used in our work at Futures Academy.

NEW DIRECTIONS FOR YOUTH DEVELOPMENT • DOI: 10.1002/yd

An overview of the history of university-community partnerships

Today most universities view civic engagement and the development of university-community partnerships as critical components of university life and culture, although the detached, elite-centered ivory tower model still dominates. University involvement in the affairs of its host community is not a new concept, but its history is a checkered one that has evolved through three distinct periods: the late nineteenth century to World War I, the post–World War I era to the assassination of Martin Luther King Jr., and 1968 to the present.

The first period: The late nineteenth century to World War I

The origin of university and community partnerships dates back to 1862, when the U.S. Congress enacted the Morrill Act. This legislation led to the creation of a cadre of universities whose mandates were focused on providing access to higher education for the working class, producing and disseminating knowledge and information to help agricultural communities, and establishing extension programs to provide technical assistance to farmers.[9]

Desirous of using a similar model for urban-based universities to address problems of the city, Daniel Coit Gilman, in his 1876 inaugural address as president of Johns Hopkins University, America's first modern research university, expressed the hope that universities would "make for less misery among the poor, less ignorance in the schools, less bigotry in the temple, less suffering in the hospital, less fraud in business, less folly in politics."[10] Other university presidents in the late nineteenth and early twentieth centuries expressed the same desire to advance knowledge to improve the quality of urban life, especially among immigrants and the poor.[11] This animating mission is found especially in the histories of four of the leading universities at the turn of the twentieth century: Johns Hopkins, Columbia, the University of Pennsylvania, and the University of Chicago.

In varied ways, these four institutions were leaders in creating an academic environment that encouraged the involvement of faculty and students in the struggle to improve the lives of residents of urban slums. Seth Low, president of Columbia University, devoted the entirety of his 1890 inaugural address to discussing the significance of the interactive relationship between New York City and Columbia and stressed the importance of faculty and students using their talents and skills to solve the problems of the city.[12]

A staunch believer in liberal education, Low nevertheless stressed the critical nexus between theory and practice. "The real world is not found in books," he said, but in cities, which are "peopled by men and women of living flesh."[13] Within this context, Low argued that scholars must be "men who see humanity, as in a vision, ever beckoning to them from behind their books. . . . The scholar without this vision is a pedant. He mistakes learning as an end in itself, instead of a weapon in a wise man's hands," a weapon, Low mused, that could be used to attack the complicated problems facing American cities.[14]

University of Chicago president William Rainey Harper went even further in formulating his vision of university-community relations. Harper believed that the university could play a leading role in transforming the United States into a socially just and democratic society. The central mission of the university, he said, was to help build a truly democratic society by taking responsibility for the performance of the entire school system within its community. He argued that "through the school system every family in the entire broad land of ours is brought into touch with the university; for from it proceeds the teachers."[15]

Harper's viewpoint was based on the notion that neighborhoods were the basic unit for urban development and that schools functioned as the hub around which neighborhood life evolved. Against this theoretical backdrop, Harper created an academic environment that nurtured the pioneering work of John Dewey, who dreamed of transforming the United States into a genuine participatory democracy by turning schools into democratic, problem-solving institutions that collaborated with residents to solve community problems.[16] Dewey believed that school-community partnerships were needed to transform neighborhoods into democratic communities imbued with the principles of reciprocity, collaboration, cosmopolitanism, and social justice. Ultimately the development of such neighborhoods would lead to the emergence of a worldwide, organic "Great Community" composed of truly participatory, democratic, collaborative, and interdependent societies. This was Dewey's dream.[17]

World War I bought this period in the history of university-community partnerships to an end. In retrospect, the visions of Low, Harper, and Dewey were ahead of their time; however, this period should not be romanticized. The university-community partnerships in this era were neither participatory nor democratic.[18] Rather, they were based on the client model of operating, in which ordinary people were viewed as consumers of the services that university experts provided.[19] The goal was to help the disadvantaged, but not to enlist them as agents of change: participants with whom they worked collaboratively to turn distressed neighborhoods into socially functional places based on participatory democracy, reciprocity, cosmopolitanism, and social justice.[20]

The second period: Post–World War I era to the assassination of Martin Luther King Jr.

During the second period, from the end of World War I to the assassination of Martin Luther King Jr. in 1968, the university-community partnership was reframed as higher education institutions redefined their public mission and their view of the problem of the city. During the late nineteenth and early twentieth centuries, universities were concerned about the plight of the urban poor because unsanitary, unsafe, and deplorable living conditions directly threatened the city's growth and development, as attested by the Great Chicago Fire in 1871. In those days, most metropolitan residents lived in highly congested central cities, and disease and violence could spread quickly from one neighborhood to another.[21]

NEW DIRECTIONS FOR YOUTH DEVELOPMENT • DOI: 10.1002/yd

This changed after World War I when the mechanization of agriculture combined with the growth of industrialization to usher in a new period of urban development. As the urban population exploded, businesses and people began moving to the suburbs, automobile traffic intensified, and universities built partnerships with community elites to construct the modern, economically rational city. In this new setting, scholars turned their attention to the problems of city and regional planning, ending the Great Depression, eliminating policy barriers to the creation of mass home ownership, and rethinking the role of higher education.[22]

In this new American metropolis, consumerism and market-based individualism triumphed as materialism defined the "good life," and the middle classes began their long trek from the central city to the suburbs. In response, scholars, led by the University of Chicago school of sociology, sought to construct a theoretical framework to justify the new approaches to city building and the restructuring of the social geography of the urban metropolis. Robert Park and Ernest W. Burgess, for example, argued that the distress found in the urban core was "a product of natural forces" and that people would move out of decaying "natural areas," as their economic conditions improved.[23] In this increasingly privatized environment, *university-community partnerships* increasingly meant collaboration with business, civic elites, and the federal government.

From the time of the Great Depression to the postwar years, the university gradually shifted its focus from local to national and international issues as the realities of war made foreign relations and national security matters of great importance. This trend was accelerated in 1945 when Vannevar Bush's report to President Theodore Roosevelt, "Science and the Endless Frontier," led to the development of a unique partnership between the federal government and the university. Bush's report called for the formation of an interactive affiliation between the federal government and colleges and universities. To accelerate the rise of the United States to international leadership and make the world safe for democracy, colleges and universities were called on to expand the frontiers of scientific knowledge. The government would aid in this process by dramatically increasing its investments in pure research, especially in medicine and the basic sciences.[24]

The Bush doctrine not only caused federal funding to turn research in science and technology into the engine that drove the massive expansion of the post–World War II university; it also enshrined the elitist, Platonic dictum, which placed "pure" over "applied" research and pushed local issues and the urgent problems facing immigrants, blacks, the poor, and working classes to the margins of academic life.[25]

The third period: The assassination of Martin Luther King Jr. to the present

The urban violence of the 1960s, which culminated with the assassination of Martin Luther King Jr. in 1968, ended the second period by forcing institutions of higher education to refocus their attention on the problems of the city. Following King's murder, violence erupted across the country as angry blacks lashed out at a society that allowed an assassin to kill the nonviolent preacher.[26] To restore hope among African Americans, predominantly white institutions of higher education opened their doors to blacks and other people of color. Black student demands for a more relevant curriculum combined with student unrest and protests over the Vietnam War to ignite the process of transforming the ivory tower into a more civically engaged university.[27]

By 1989, the ending of the cold war combined with these internal changes and a deepening of the urban crisis to pave the way for the development of a new generation of university-community partnerships. From the 1960s onward, the condition of the cities continued to deteriorate. In 1965, when the black scholar Kenneth B. Clark referred to Harlem as a *dark ghetto*, he was talking about the emergence of the inner-city built environment as the epicenter of racism, structural inequality, joblessness, poverty, underperforming schools, dilapidation, family instability, crime, and violence.[28]

These issues stood at the doorstep of universities and forced them to become genuinely concerned about the problems of distressed urban communities and their poor and working-class residents.[29] As a consequence, since the late 1980s, civic engagement and the development of partnerships with its host community have become an acceptable practice in most universities as they became more democratic, people centered, and cosmopolitan. We are still in the very early stages of this transformational process and still learning how to construct a university-based model of civic engagement that turns schools into democratic problem-solving institutions that collaborate with residents and stakeholders to solve community problems.

In such a model, the university assists in the establishment of a school-centered model of community development that links schooling to community building and neighborhood transformation. In this way, young people will began to see the connection between the lessons learned in school and their ability to work with neighbors and stakeholders to build a better community. Thus, by involving young people in a democratic and collaborative process to transform their community, we will turn them into good students, who will become caring, productive, and engaged citizens.

The remainder of this article focuses on our efforts to contribute to the development of such a university-assisted model of school-centered community development. However, before discussing the Futures Academy experience, we provide an overview of the key learning theories used in the construction of our pedagogic model. Then, using Futures Academy as the engine that drives the remaking of the Fruit Belt neighborhood, we illustrate our quest to develop young people by meaningfully involving them in the quest to turn their community into a socially functional neighborhood characterized by participatory democracy, reciprocity, and collaboration.

The pedagogical model: Problem-based learning, youth development, and community building

Linking students and their schools to the community development process requires the evolution of a theory of learning and instructional strategy capable of developing students' critical thinking abilities. Such a task demands the transformation of both teacher and student. Within this context, at the same time that universities begin to forge authentic partnerships with distressed communities, it is critical to also create an environment that encourages scholars to develop pedagogical approaches to grapple with this issue. In this sense, the creation of pedagogic approaches that will lead to authentic learning and transform the culture of schools is a critical first task in the construction of a school-centered model of community development that makes youth development its focal point.

It is natural to begin with John Dewey, who theorized that education and society are interactive and interdependent. Thus, the only way to build a society based on participatory democracy is to construct an effective democratic schooling system, one informed by a pedagogic approach capable of turning young people into critical thinkers who are caring, productive, and civically engaged citizens. At the core of this endeavor is the question, "How do you create a democratic classroom where students become critical thinkers and problem solvers, imbued with the values of reciprocity, collaboration, cosmopolitanism, and social justice?"

We based the development of our pedagogic model on a synthesis of the work of John Dewey, Paulo Freire, and other theorists of active learning. Dewey and his concept of the Great Community provide a democratic education model based on "the very same ideals that inspired the Declaration of Independence . . . those of democracy, of the liberty and equality that animated our forefathers."[30] Action is a core principle in the Dewey philosophy, and his notion of democracy is rooted in the ideal of racial, social, and economic justice. It is conceived as a robust, interactive way of life in which students, on their way to becoming participatory citizens, are continually engaged in the quest to solve complicated neighborhood and societal issues. This is conceived as an interactive process of problem solving that will continually recreate and re-form society.

Dewey's Lab School was an effort to activate these ideals into a reproducible educational model, but it did not result in an authentic process that translated Dewey's great ideas into action for students.

The programs were child focused, involved hands-on activities, and fostered problem solving, but they were implemented within the existing institutional framework of academia, where transformative projects and real-world knowledge are not core values.[31]

Paulo Freire's pedagogical model in many ways builds on Dewey's ideas. Formed by the concepts of dialogic education and praxis, his approach to education prepares students to "analyze social life through a lens of diversity and social justice and . . . be transformative social agents."[32] Dialogic education honors the knowledge and experience of both students and teachers and seeks to build on both. Central to the learning process is the awareness that unequal power relationships exist and that an important goal of transformative education is to give voice to the silenced, while also investigating the cause of that silence, thereby unlocking their critical consciousness and creative powers.

Praxis involves both action and reflection in a looping fashion, with one ever leading to the other. One cannot obtain critical consciousness by focusing only on intellectual pursuits: "reflection, both self and social, coupled with dialogue can foster a critical consciousness by which students and teachers see their experiences situated in historical, cultural, and social contexts and recognize possibilities for changing oppressive structures."[33] This educational model views student and teacher as equal actors in the learning process, which is ultimately tied to action and transformation of community and the rest of society.

David A. Kolb's active learning cognition theory also conforms closely to the basic tenets of Dewey, Freire, and critical pedagogy. He emphasizes the importance of bringing lived experiences into the classroom for reflection and believes that simulations and case studies, coupled with lectures and reading, would round out the learning process and tie action to reflection. Kolb hypothesizes that there are four stages of learning: concrete experience, observation and reflection, the formation of abstract concepts, and testing in new situations. These stages create a continuous cycle that can be entered into at any point but must be followed sequentially to create an engaged learning environment. The permeable boundaries between the classroom and the outside world are thought to enhance transmission of key knowledge to the larger community and vice versa.[34]

Problem-based learning (PBL) is also a method developed to produce engaged, active learners. Students are responsible for their learning, which involves searching for solutions to issues that occur in the real world. "PBL is focused, experiential learning organized around the investigation, explanation, and resolution of meaningful problems."[35] Like Kolb's theory, PBL has a learning cycle composed of stages to be followed sequentially. The student begins with a problem, real or hypothetical, identifies the key facts, and then generates a hypothesis. The self-directed aspect comes into play as the student, in trying to resolve the problem, identifies deficiencies of knowledge in the next stage. The work is then to acquire and apply new knowledge in the next phase. The last stage is abstraction—reflection on the total problem-solving process.

Throughout this process, students continually negotiate with other students and teachers in a cooperative, collaborative fashion to test out old and construct new categories of knowledge. Teachers serve as guides moving students through the cycle, and each should be "an expert learner, able to model good strategies for learning and thinking, rather than an expert in the content itself."[36] Each problem-solving cycle is intended to further each student's understanding of a self-determined goal that has been set; problem resolution is not an end in itself. The idea is to further the development of metacognitive skills in the students. These involve the ability to plan one's own problem-solving process, as well as to monitor and evaluate it.[37]

Engaged, action-based learning and reflection is the core concept that links together these various theories of learning. The guiding principle is that real-life issues provide opportunities for teachers and students to collaborate, problem-solve, and reflect, and this leads to the formation of critical consciousness and authentic participatory democracy. This approach reinforces Dewey's notion that the intelligence and maturity of children develop best when they are involved in the quest to solve the puzzling real-world

problems confronting them and their families and given the opportunity to reflect deeply on these problems.[38]

The key to the development of a pedagogical model based on
these theories of engaged learning is to structure practical activities that enable students to use the knowledge and skills they learn
in the classroom to reflect on neighborhood problems and work
collaboratively with residents and stakeholders to solve them.
Against this backdrop, we developed a neodemocratic education
model to guide our work at Futures Academy.[39] In this approach,
the goal is not simply to turn young people into good students,
equipped with the knowledge and skills required to earn a living.
It is also to imbue them with the desire to build a better, more
socially just world.

This type of pedagogical method is critical in an inner-city setting, where so many students underperform academically, drop out
of school, and make poor choices that sometimes lead to premature death or incarceration. This happens, we argue, because inner-
city students do not see a relationship between education and the
ability to improve their lives and make their neighborhood a better place to live. We believe that unless students understand this
vital connection between education and community building, they
will not be motivated to learn and develop their talents and skills
fully.[40] Thus, our pedagogic model is not only a method of teaching; it is also a community-building activity that contributes to the
holistic development of young people: good students, engaged
neighborhood residents, and community change agents.

Connecting schooling, youth development, and commu-
nity building: Futures Academy

Futures Academy is a struggling public school for kindergarten to
eighth grade, located in one of the poorest neighborhoods in Buffalo, New York. At the same time, because it is situated adjacent to
the Buffalo-Niagara Medical Campus, where the State University
of New York at Buffalo has a strong presence, it is an ideal site to

launch a university-assisted school-centered community development project.

Futures Academy is a neighborhood magnet school that draws
its students from inside the neighborhood and across the city.
Although originally meant to be a magnet school that offered students a curriculum to prepare them for careers likely to be important in the future, Futures now uses its magnet school status only
as a vehicle for recruiting students citywide. About a third of the
694 students at Futures come from the Fruit Belt, with the remainder being drawn from other low-income neighborhoods in Buffalo.

The school is predominantly African American, with a handful
of whites, Latinos, and Native Americans. All students attending
Futures are eligible for free or reduced-price school lunches, and
the school performs well below New York standards at all grade
levels in English language arts and math. Most of the teachers at
Futures have more than three years of experience, and about 19
percent have a master's degree or doctorate.[41] The school is headed
by a progressive African American principal with a doctorate who
grew up in the Fruit Belt neighborhood.[42]

Our school-centered development project is an initiative we call
the Community Classroom Program. Administered by the Center
for Urban Studies (CENTER) at the State University of New York
at Buffalo, the program involves most sixth through eighth graders
in activities both during and after school hours. The Community
Classroom complements the school's curriculum but is not integrated into regular classroom activities. Rather, during the school
day, students participating in the program are given release time
from their classroom, and for after-school activities, they are
required to obtain permission from their parents. Graduate students from the university drive the program, but a number of classroom teachers assist in the development of all program activities,
including those that take place after school.

The Community Classroom Program uses the Fruit Belt neighborhood as a classroom and engages students in collaborative activities with residents to solve community problems. The goal is to
create opportunities for the students to apply the knowledge and

skills they have learned in the classroom to the goal of making the Fruit Belt a better place to live by working in collaboration with residents and stakeholders. The program seeks to implement the Dewey dictum that individuals learn best when they are involved in the quest to solve the puzzling real-world problems confronting them and their families and when they are given the opportunity to reflect deeply on these problems. The Community Classroom consists of four interrelated activities: Future City Project, Community Clean-A-Thon, Community Garden Project, and the Community Art Program.

Future City Project

The goal of the Future City Project is to show students that a connection exists between public policy and the city and neighborhood development process. The idea is to debunk the notion that conditions in their neighborhood or elsewhere are the products of a natural developmental process rather than the outcome of a human decision-making and resource allocation process. We want the students to understand that agency—the action of residents in partnership with other stakeholders and the government—can improve conditions in their neighborhood by altering the policies and decisions that drive community development.

The Future City Project is a simulated problem-solving activity with real-world implications. Each year, as part of a broader national competition, we develop two to three teams of six students, composed of sixth through eighth graders, who build a futuristic city based on a specific theme, such as nanotechnology, transportation, or alternative energy sources. As part of the competition, the students, using SimCity software, also develop a computerized city and then construct a scale model of a smaller portion of it. In this process, they explore various policy choices and decide which ones to apply in the building of their city. The students take field trips to deepen their understanding of the theme and gain insight into ways that neighborhoods and cities are shaped by policymaking and decision-making processes. Local engineers and urban planners are always enlisted to work with the students in developing their project.

NEW DIRECTIONS FOR YOUTH DEVELOPMENT • DOI: 10.1002/yd

Between September and January, the students construct their computer city and a scaled model of a smaller section of it. After the January competition, the students are required to reflect on their experiences: they engage in group discussions about lessons learned and write a short essay on their experiences. After the reflection exercise, they spend the remainder of the school year working on select neighborhood projects. The idea is for them to use the knowledge and skills they learned in the Future City competition to work on real-life problems in their own neighborhood.

The Community Art Project

The Community Art Project involves students in the struggle to change the visual image of their community through art. The principle is to show students how they can change the way their neighborhood looks and feels. Dilapidation and a forlorn environment do not have to be the characteristic features of distressed communities. Within this framework, we want students to think aggressively about ways to reimage their community and imbue it with the energy of youth culture. Over the past five years, the students have produced some rather dramatic projects, such as working in partnership with the Locust Street Neighborhood Art Classes, Inc., which is a nonprofit organization that provides free art and photography classes for young people. They produced a mural of about four hundred small panels to cover the fence surrounding a small neighborhood park. They also designed and built two benches for the park.

The students produced a unique sign, which consisted of a bench and a decorative archway, for a block-long garden/park designed by Futures students and built by the university's Center for Urban Studies. Moreover, while the Futures Academy school building was being rehabilitated, the students were permitted to develop a mural along the wall fronting the entrance to the school. The mural consists of several hundred small tiles, each painted with a different design. Now, the first thing they see when entering the school is the mural, which symbolically proclaims, "This school belongs to you." And the first thing they see when they leave school is the sign and garden they designed, which symbolically says to them, "This neighborhood belongs to you."

The students have also developed art projects designed to get young people to "stop the violence" and turn derelict old houses into works of art. In the housing project, the students used 4 × 8 plywood boards as canvasses for their paintings, which were then installed over the doors and windows of dilapidated houses that had been scheduled for demolition. When the house is demolished, the panels will be removed and placed on another structure. The project demonstrated that it is possible to use youth art to change the visual image of structures that had been community eyesores.

The public spaces, on which the community art projects have been erected, have become "sacred" places and are never vandalized. Thus, the work of the students is becoming a real part of their community, not only increasing the aesthetic value of the environment but sending positive, uplifting messages to all who live and work there. This is a real sign of active citizenship.

The Community Garden Project

The goal of the Community Garden Project is to solve the problem of unkempt vacant properties in the Fruit Belt. The project centers on two main activities. First, the students at Futures Academy are involved in the ongoing maintenance and development of the Futures Garden, a passive park that fronts the school. A passive park is one that is designed to encourage meditation, picnicking, walking, playing, and observation of flowers and community art. Four years ago, the garden site was a series of unkempt vacant lots that symbolized the powerlessness of the students, teachers, and neighborhood residents. To the children, these lots seemed to say, "You are not worth much, and no one really cares."

Futures students, in partnership with neighborhood residents and the Center for Urban Studies, decided to turn this message around. Graduate students assisted the students in planning a passive garden, acquiring control over the land, and overseeing the physical development of the park. The students learned that even with limited resources, they had the power to alter the visual image of the community through a vacant lot management strategy. Today their task is to maintain and further develop the Futures Garden.

Second, the students are involved in the development of a model vegetable garden with neighborhood residents. Here, they are learning about urban gardening, nutrition, and healthy meals. As part of this project, the children were involved in a bioremediation project in which they learned how to use plants to cleanse the soil of contaminants.

The Community Clean-A-Thon

While in the Future City Project, the students are required to solve a simulated problem. In the Community Clean-A-Thon, they are solving a real-world problem. The Community Clean-A-Thon is a year-long program that involves students in overcoming neighborhood blight. In the fall 2008 program, students analyzed the pattern of rubbish and trash dumping in the community and formulated a strategy for solving this problem.

Between September and December, the students studied the distribution of trash and rubbish in the Fruit Belt. They completed two main tasks during this period: they identified the location of clusters of rubbish (old tires, discarded appliances, bottles, and so forth), and they examined the distribution pattern of the clusters of rubbish and trash. The students used the GIS (Geographic Information Systems) method to analyze the distribution pattern. Graduate students in the Department of Urban and Regional Planning teach the students about GIS and show them how to geocode, map, and analyze the data. In the winter and spring, the students completed their analysis of the distribution pattern, generated hypotheses on causality, and formulated a plan for solving the problem. In the fall of 2009, the students will implement and evaluate the plan for solving neighborhood blight.

At the end of the school year, Futures Academy holds a Community Clean-A-Thon that draws the entire community and stakeholders into a neighborhood cleanup. The Clean-A-Thon is organized around the theme, "Collective Work and Responsibility," which stresses the importance of the entire community taking control of the development of their neighborhood. The students, who have been working on the project throughout the year, now join with other teachers, residents, and stakeholders to plan the Clean-A-Thon. Based on the GIS mapping project, the committee develops a plan for deploying the participants in cleanup activities throughout the neighborhood.

The morning hours of the Clean-A-Thon are devoted to cleaning up the neighborhood, and the afternoon is set aside for a community celebration. Thus, the morning hours are about work, while the afternoon focuses on a neighborhood coming together to feast and have fun. Most important, the festival creates an opportunity to deepen the bonds betwixt and between teachers, students, residents, and stakeholders.

Lessons learned and challenges ahead

The principal at Futures Academy says that students enrolled in the Community Classroom Program are not only doing well in their classes, but also are developing into mature youngsters who try to keep their friends out of trouble. Thus, our experiences reinforce the belief that we can turn young people into good students, who will become caring, productive, and engaged citizens, by involving them in a democratic and collaborative process to improve their neighborhoods. Within this context, the most important lesson learned is that part of the task of creating a democratic classroom consists of getting the students out of the school building and into the community, where they participate in collaborative neighborhood problem-solving activities.

Learning activities in the school building, no matter how creative and thoughtful, are limited in their ability to get students to see the critical nexus between schooling and community development. The only way for them to see this connection is by participating in problem-solving activities that take place in the community. Consequently, every university-assisted school should have a community classroom program that involves students in community problem-solving activities with residents and stakeholders. Only by embedding students in the community change process can we develop young people who are critical thinkers and problem solvers, imbued with the values of reciprocity, collaboration, cosmopolitanism, and social justice.

The great challenge we face in making this happen is to turn universities and public schools, both autocratic institutions, into truly democratic places that believe in the transforming power of critical thinking and participatory democracy. This is a big task that must start with encouraging the widespread study and discussion of the meaning of democracy. On this point, we stress that the study and discussion of democracy cannot occur apart from practice. Thus, one of the most important, and difficult, challenges we face is how to create activities whereby people learn about participatory democracy through the process of building authentic democratic institutions. This is one of the keys to expanding university-assisted school-centered community development programs that make youth development the focal point of activities.

Notes

1. Green, J. P. (2001). *High school graduation rates in the United States.* New York: Manhattan Institute for Policy Research, Center for Civic Improvement.

2. Johns Hopkins Researchers. (2007, Oct. 29). Dropout factories: Take a close look at failing schools across the country. Associated Press. Retrieved December 4, 2008, from http://www.ferndaleschools.org/parentresources/documents/Det%20News_Oct29_DropoutFactories_article.pdf.

3. Butterfield, E. (2007, April 7). Prison rates among blacks reach a peak, report finds. *New York Times.* Retrieved September 9, 2008, from http://www.nytimes.com/2003/04/07/us/prison-rates-among-blacks-reach-a-peak-report-finds.html.

4. Amnesty International. (2003). *Human rights Watch World Report: United States.* Retrieved November 8, 2008, from http://www.hrw.org/legacy/wr2k3/us.html.

5. Crane, J. (1991). The epidemic theory of ghettos and neighborhood effects on dropping out and teenage childbearing. *American Journal of Sociology, 96,* 1226–1259.

6. Harkavy, I. (1999). School-community-university partnerships: Effectively integrating community building and education reform. *Universities and Community Schools, 6*(1–2), 7–24.

7. Caughy, M. O., Nettles, S. M., & O'Campos, P.J.P. (2008). The effect of residential neighborhood on child behavior problems in first grade. *American Journal of Community Psychology, 42*, 39–50.

8. Anderson, G. L. (1009). Toward authentic participation: Deconstructing the discourses of participator reforms in education. *American Educational Research Journal, 35*, 571–603.

9. LaMay, C. L. (2001). Justin Smith Morrill and the politics of the landgrant college acts. In L. K. Grossman & N. N. Minnow (Eds.), *A digital gift to the nation: Fulfilling the promise of the digital and information age* (pp. 73–95). Washington, DC: Century Foundation.

10. Harkavy. (1998).

11. Benson, L., & Harkavy, I. (2000). Higher education's third revolution: The emergence of the democratic cosmopolitan civic university: *Cityscape, 5*, 47–57.

12. *Proceedings of the installation of Seth Low, LL.D., as president of Columbia College in the City of New York.* (1890, February 3). New York: Columbia College.

13. *Proceedings of the installation of Seth Low.*

14. *Proceedings of the installation of Seth Low.*

15. Benson, L., Harkavy, I., & Puckett, J. (2007). *Dewey's dream: Universities and democracies in an age of education reform.* Philadelphia: Temple University Press.

16. Benson et al. (2007).

17. Benson et al. (2007). p. ix.

18. Harkavy. (1998).

19. Hess, D. J., Lanig, H., & Vaughan, W. (2007). Educating for equity and social justice: A conceptual model for cultural engagement. *Multicultural Perspectives, 9*, 32–39.

20. Hess et al. (2007).

21. Taylor, H. L. Jr. (2000). Creating the metropolis in black and white: Black suburbanization and the planning movement in Cincinnati, 1900–1950. In H. K. Taylor Jr. & W. Hill (Eds.), *Historical roots of the urban crisis: African Americans in the industrial city, 1900–1950* (pp. 51–71). New York: Garland.

22. Taylor. (2000).

23. Burgess, E. W. (1925). *The urban community.* Chicago: University of Chicago Press. p. 5.

24. Bush, V. (1945). *Science: The endless frontier—A report to the president.* Washington, DC: U.S. Government Printing Office.

25. Harkavy. (1998).

26. Gale, D. E. (1996). *Understanding urban unrest: From Reverend King to Rodney King.* Thousand Oaks, CA: Sage.

27. Hine, D. C. (1992). The black studies movement: Afrocentric-traditionalist-feminist paradigms for the next stage. *Black Scholar, 22*, 1–11; Joseph, P. E. (2003). Dashikis and democracy: Black studies, student activism, and the black power movement. *Journal of African American History, 88*, 182–203.

28. Clark, K. B. (1965). *Dark ghetto: Dilemmas of social power.* New York: HarperCollins.

29. Benson et al. (2007).
30. Dewey, J. (1936, January 14). Education and the new social ideals. *Vital Speeches of the Day, 2*(11), 327.
31. Benson et al. (2007).
32. Nagda, B., Gurin, P., & Lopez, G. (2003). Transformative pedagogy for democracy and social justice. *Race, Ethnicity and Education, 6*, 165–191.
33. Nagda et al. (2003).
34. Kolb, D. A. (1984). *Experiential learning: Experience as the source of learning and development.* Upper Saddle River, NJ: Prentice Hall.
35. Hmelo-Silver, C. (2004). Problem-based learning: What and how do students learn? *Educational Psychology Review, 16*, 235–266. Quotation is from p. 236.
36. Hmelo-Silver. (2004).
37. Sperling, R., Howard, B., Miller, L., & Murphy, C. (2002). Measures of children's knowledge and regulation of cognition. *Contemporary Educational Psychology, 27*, 51–79; Hmelo-Silver. (2004).
38. Benson et al. (2007).
39. Forde, S. (1996). Review: *Rethinking democratic education: The politics of reform*, by David M. Steiner. *Journal of Politics, 58*, 270–271.
40. Taylor, H. L. Jr. (2005). Connecting community development and urban school reform. In M. E. Finn, L. Johnson, & R. Lewis (Eds.), *Urban education with an attitude* (pp. 41–57). Albany: State University of New York Press.
41. P.S. 37 Futures Academy, New York. (2009). *Great schools.* Retrieved September 9, 2008, from http://www.greatschools.net/modperl/browse_school/ny/373.
42. The partnership with Futures Academy is part of a broader neighborhood development initiative led by the Center for Urban Studies and the Community Action Organization of Erie Academy. Futures Academy is situated in the Fruit Belt, a small community of about three thousand residents, with the demographic profile characteristics of a distressed urban neighborhood: low incomes, high poverty rate, high unemployment, and underemployment, combined with crime and a proliferation of single-parent families and a weak organizational structure. The physical environment is characterized by dilapidation, blight, and vacant lots. Although it is one of the poorest neighborhoods in the region, the Fruit Belt nevertheless has considerable assets. It is home, for example, to the Buffalo-Niagara Medical campus and St. John Baptist Church. The medical campus contains the region's top clinical, research, and medical education institutions, and the church is the largest and most powerful black church in western New York State.

HENRY LOUIS TAYLOR JR. *is the founding director of the Center for Urban Studies and a full professor in the Department of Urban and Regional Planning at the State University of New York at Buffalo.*

LINDA GREENOUGH MCGLYNN *has a Ph.D. in social welfare and is in private practice.*

Community partnerships are powerful in direct proportion to the strength of the relationships that have been forged. The partnership network that has been devised among Indiana University-Purdue University Indianapolis, the George Washington Community High School, and the local community defines the relational qualities needed to make sustainable changes within the high school.

3

George Washington Community High School: Analysis of a partnership network

Robert G. Bringle, Starla D. H. Officer, Jim Grim, Julie A. Hatcher

INCREASINGLY COLLEGES AND universities are becoming more civically engaged in their local communities. This community engagement is recognized as a valuable dimension of higher education fulfilling its civic role. In addition, this engagement holds potential for improving teaching and learning, as well as scholarship and research. Yet understanding the benefits of such engagement to communities is an underresearched area within the domain of engagement. Two major areas can be studied in terms of understanding the value of civic engagement for communities: (1) changes in outcome measures that are associated with quality

NEW DIRECTIONS FOR YOUTH DEVELOPMENT, NO. 122, SUMMER 2009 © WILEY PERIODICALS, INC.
Published online in Wiley InterScience (www.interscience.wiley.com) • DOI: 10.1002/yd.305

41

of life in communities, such as a decrease in teenage birth rates or an increase in reading scores, and (2) the relationships that emerge and develop between community and campus.

Typically this relationship is described as "campus-community," "community-campus," or "university-school" in the case of higher education and K–12 schools. Each of these phrases assumes that two entities comprise the relationship. From our collaborative work over the past decade to improve educational opportunities in our community, we now recognize and value the strength and rewards of a network of relationships.

In this article, we expand the notion of the campus-community partnership perspective by describing a network of relationships between and among four stakeholders: the university, the school, community organizations, and residents. We then analyze the unique partnership between Indiana University-Purdue University Indianapolis (IUPUI) and George Washington Community High School (GWCHS) in terms of how it has developed qualities of a relationship that are desirable in civic engagement work.

Community-campus partnerships

One way of assessing civic engagement is for a campus to count the number of campus-community programs that exist. As important as this may be for benchmarking the status of civic engagement, it is vital to shift the focus from the quantity to the quality of such relationships. Developing better campus-community relationships is viewed as one of the basic building blocks for universities to improve civic engagement work by universities.[1] Furthermore, Cruz and Giles recommend as a remedy for the paucity of community-focused research "that the university-community partnerships itself be the unit of analysis."[2] But how can these relationships, specifically between a university and a K–12 school, be studied, and how can the quality of the relationship be assessed?

NEW DIRECTIONS FOR YOUTH DEVELOPMENT • DOI: 10.1002/yd

At the most general level, we propose that four clusters of constituencies can be differentiated for analyzing university-school relationships: the school, the university, the residents of the community, and community organizations, such as businesses, nonprofits, and government agencies. Each of these entities plays an important role in transforming schools to improve the academic success of youth. Accordingly, there are six relationships that exist among these four constituencies (Figure 3.1).

The term *partnership* is sometimes used in the most generic sense to describe interactions between entities; it is also used to denote interactions that possess particular qualities. At a national Partnership Forum, convened by Portland State University in March 2008, representatives of community organizations and higher education discussed the term *partnership*.[3] The ideas proffered were summarized in the following statement:

Partnerships develop out of relationships and result in mutual transformation and cooperation between parties. They are motivated by a desire

Figure 3.1. Network of six relationships

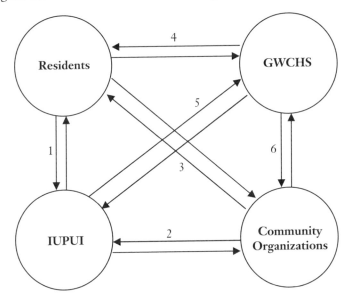

to combine forces that address their own best interest and ideally result in outcomes greater than any one organization could achieve alone. They create a sense of shared purpose that serves the common good.[4]

For this analysis, we use the term *relationship* to refer to personal interactions between people and *partnership* to describe a particular subset of relationships that is characterized by three qualities: closeness, equity, and integrity (see Figure 3.2). Thus, a campus or a school may have a number of relationships with outside entities, but there may be only a limited number of partnerships because not all relationships become partnerships.

History of the GWCHS-IUPUI partnership

Only a narrow river separates the IUPUI campus from the Near Westside Indianapolis community. However, the bridges that connect this urban research campus of more than thirty thousand students with its neighbors to the west are both real and symbolic. Relationships between IUPUI and the Near Westside have taken years to develop, yet their strength gives support for a number of

Figure 3.2. Types of relationships

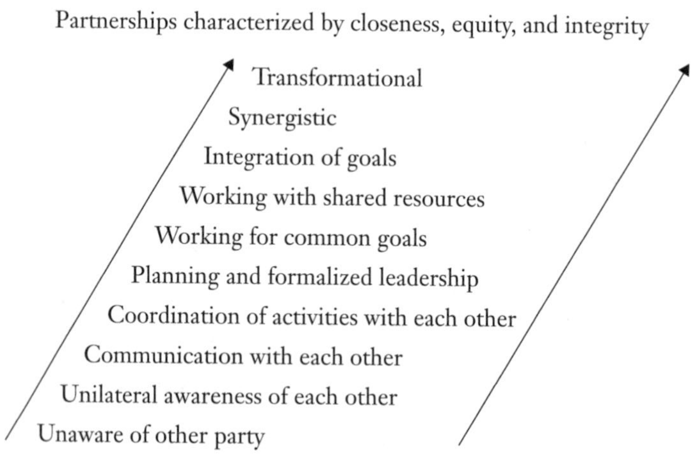

Partnerships characterized by closeness, equity, and integrity

Transformational
Synergistic
Integration of goals
Working with shared resources
Working for common goals
Planning and formalized leadership
Coordination of activities with each other
Communication with each other
Unilateral awareness of each other
Unaware of other party

faculty, staff, students, and community residents to cross back and forth between the campus and community. This collaboration is best illustrated at GWCHS, a public school within a mile of the campus. The very existence of GWCHS is an achievement of this campus-community partnership.

Historically, George Washington High School had reflected the educational and civic heart of the Near Westside of Indianapolis. Residents of this largely working-class area, comprising three distinct and diverse neighborhoods, took pride in the historical, and often family, legacy of the school. The school was closed in 1995, as were all elementary schools in the area, as the Indianapolis Public School (IPS) system was forced to consolidate schools as enrollment fell. The residents united to fight against the closure but lost their case before the IPS School Board.

At about the same time, the IUPUI Office of Neighborhood Partnerships (ONP) was established by the campus to build long-term strategic partnerships with the Near Westside community. Through early conversations and an asset-mapping of the community in 1996, ONP and community representatives identified education of youth as one target for joint activity. Due in part to activities for a Department of Housing and Urban Development Community Outreach Partnership Center (HUD-COPC) grant, a public forum convened community residents, staff from ONP, and community organizations. The highest priority that was expressed in forging the campus-community collaboration was reestablishing public schools in the Near Westside. The goal would be to focus initially on secondary schools and then on elementary schools.

IUPUI, residents, and community organizations

The grassroots activities to reestablish public schools on the Near Westside created new relationships between the campus, residents, and community organizations that had not previously existed (see relationships 1, 2, and 3 in Figure 3.1). As a result of the forum, residents of the community, representatives from community organizations, and faculty and staff from IUPUI formed the Westside Education Task Force (WETF). Since its inception in 1996, the WETF has organized forums, coordinated study circles, conducted surveys of educational needs and interests of residents, and collaborated with IPS staff and the superintendent to identify ways to improve educational opportunities for residents in the Near Westside. Facilitated by ONP, these activities have brought community residents, representatives from community agencies, and IPS staff together to create a vision for a community-oriented school in the neighborhoods.

Thus, the WETF became a formal structure for establishing relationships among IUPUI, the residents, and staff from community organizations before any school existed in the neighborhood. After a team of representatives visited schools in Louisville, Boston, and New York City, the WETF proposed that any school that reopened in the Near Westside neighborhoods should have the capacity to be an integral part of the community by providing a broad range of services for students, their families, and the larger community. WETF members thought this could be accomplished most successfully through implementation of a community school model. This model promoted strong partnerships between the school, social service providers, parents, and the community to provide a network of services and for the schools to become the center of community life.

Before George Washington reopened in fall 2000, the principal, teachers, and other educators met with parents, neighborhood leaders, community organizations, university staff and faculty, service providers, and faith-based representatives to plan how a community-focused school would operate. Parents and community residents played a key role in defining this, and they modeled a collaborative, participatory, and democratic approach to decision making. Although George Washington originally opened as a middle school, high school grades were added following the 2000–2001 school year, one year at a time. As a result, the school officially changed its name to George Washington Community High School. In 2006, Wendell

Phillips School 63 was opened in the Near Westside as a community elementary school.

The school and the university

For the past eight years, GWCHS has collaborated with IUPUI to provide mutually beneficial relationships and programs for GWCHS and IUPUI students, faculty, and staff (see relationship 5 in Figure 3.1). Many of these activities have been generated through the IUPUI Center for Service and Learning, a centralized unit on campus that encompasses four offices: Neighborhood Partnerships, Service Learning, Community Work-Study, and Community Service. For example, America Reads and America Counts tutors provide free tutoring in reading and math to youth through school-based and after-school programs. Each year, the Sam H. Jones Community Service Scholarship program places between fifteen and twenty-five college students in after-school programs and college readiness programs offered as part of the social studies curriculum. College students coach cheerleading, assist the school nurse, conduct fitness classes, provide tutoring, offer art classes, and serve as athletic trainers. School personnel believe that the mere presence of college students in the school building has contributed immensely to the increased rate of graduates continuing on to postsecondary education.

In addition to coordinating the HUD-COPC activities that focused on educational issues, financial literacy, and health initiatives, ONP serves as a catalyst for linking other campus units with GWCHS and the Near Westside. As active participants in community forums organized through the Great Indy Neighborhoods Initiative, ONP staff keep an up-to-date list of neighborhood priorities in education, youth involvement, health, and civic engagement. A faculty development program, Community Fellows, involves six faculty from IUPUI in a year-long faculty learning community focused on developing new community partnerships in the Near Westside. The schools of Education, Nursing, and Physical Education have developed comprehensive programs at GWCHS through preservice and service-learning classes.

Annually, approximately seventy-five IUPUI students, faculty, and staff work with GWCHS, an estimated value of more than $300,000 to the school. This work is consistent with the university's commitment to civic engagement as an integrated part of the campus mission. Involvement at GWCHS provides an opportunity for college students to receive hands-on experience with students, teachers, administrators, and other partnering agencies. College students are able to work within a thriving set of community resources in one of the most poverty-stricken communities in Indianapolis. In doing so, they have an opportunity to have an impact on the future development of the community by assisting GWCHS students and their families with education. The college students serve as role models, and as they share their stories with high school students, they provide encouragement for them to continue into postsecondary education. The experiences shape their civic commitments and confirm their career choices. The return on investment is equally valued by both partners.

GWCHS and community residents

GWCHS representatives along with ONP staff collaborated in the Near Westside Great Indy Neighborhoods Initiative, a neighborhood strategic planning activity (see relationship 4, Figure 3.1). More than 150 residents, many new to neighborhood engagement, participated in these activities. The planning resulted in several resident-driven committees working to improve neighborhood issues such as a business association, education, housing, health, and public safety. Most recently, with funding from the Indiana Campus Compact, a "Listening to Communities Dialogue" was convened by ONP staff at the neighborhood library. GWCHS students participated and contributed to conversations that will enhance neighborhood work. In partnership with IUPUI, GWCHS now offers fitness and nutrition services to students, parents, and the broader community in a newly created wellness center. Nursing and physical education students staff the wellness center, which provides low-cost access for residents. The Fit for Life fitness program now draws 145 community residents who work out in the wellness center. In addition, two centers for working families are located in the geographical area surrounding GWCHS. These centers educate the community on the importance of addressing financial literacy.

Since GWCHS opened, it has been effective in both reaching out to the community residents and inviting them into the school. By hosting community events such as family nights, meals, and health fairs, having a community meeting room available for free use, and providing access to a swimming pool and wellness center, GWCHS has reestablished itself as a hub for the neighborhood. In 2005, a survey of a hundred Near Westside households identified the schools as the greatest neighborhood asset. This is indicative of a relationship between the school and community residents that has increased in closeness, is viewed as mutually beneficial, and has integrity by demonstrating common concern for improving the neighborhood.

GWCHS and community organizations

Over the past eight years, extensive relationships have emerged between the GWCHS and community organizations (see relationship 6 in Figure 3.1 and Table 3.1). GWCHS has relationships with more than fifty businesses, organizations, neighborhood groups, and service providers (including IUPUI) that have resulted in programs and services valued at more than $2 million annually for students, their families, and residents. The presence of a thriving school has also contributed to the economic revitalization of the West Washington Street corridor. Buildings that were once boarded up are now thriving businesses ranging from banks and bakeries to health care providers and restaurants. This growth, paired with community planning, has resulted in the development

Table 3.1. Examples of GWCHS partnerships with community organizations

Community Organization	Activity
Indy Parks and Recreation	Manages the swimming pool, providing lifeguards and swimming instructors for school day physical education and after-school programming, evening and weekend public swimming and classes.
Young Audiences of Indiana	Provides a series of arts workshops for students in after-school hours that include theater, African drums, original music production, and art created from neighborhood service-learning cleanup projects.
LaPlaza Community Organization	Offers TuFuturo programming to help Hispanic students navigate college financial aid, standardized tests, and entrance applications, and weekly programming for middle school girls and their mothers, as well as boys and their fathers.
Indianapolis Urban League	Mentors ninth- and tenth-grade African American and Latino students in Project Ready to help ensure high school graduation and postsecondary learning with two on-site mentors that meet twice weekly during the school day and extended hours.
ACE Mentoring	Provides high school students project-based mentoring focused on architecture, engineering, and construction technology careers, culminating with a spring design project the student team presents to a public forum of hundreds.
Hawthorne Community Center	Coordinates activities including after-school tutoring and homework assistance, community service projects, swimming, life skills development, Peace in the Streets, and antitobacco and drug prevention.
Midtown Community Mental Health Services	Provides three full-time counselors for students and their families with access to specialist treatment and medications during school days and extended hours for the convenience of GWCHS families.
Marion County Probation Officer	On-site service for Westside youth on probation, reporting a 70 percent decrease in the GWCHS total since relocating here three years ago.
Neighborhood Centers for Working Families and the Indianapolis Neighborhood Housing Partnership	Offers parents and adults monthly workshops on family budgets, mortgage payments, purchasing and maintaining a home, and critical employment skills. Workshops are offered in conjunction with monthly GWCHS Family Nights, which include dinner, games, and prizes, free of charge, to promote parent-school engagement in a social setting.

NEW DIRECTIONS FOR YOUTH DEVELOPMENT • DOI: 10.1002/yd

of a new Westside business association. The community school model has been effective because of the investment of these many organizations.

The extent of this success is facilitated on a daily basis by a full-time community school coordinator, a professional staff position who is responsible for coordinating a diverse set of activities between the school and community organizations. For instance, Midtown Community Health provides on-site mental health counseling, which is now viewed as a seamless part of the school services. Hawthorne Community Center provides after-school programming at GWCHS to all seventh and eighth graders from the neighborhood. The Community Advisory Council provides a forum to engage and connect new and existing partners; staff and faculty from IUPUI participate on this advisory council. The school's investment in developing lasting relationships is a long-term commitment of the teachers and key administrators. These relationships provide a means for not only strengthening the community connections of GWCHS but also developing long-term partnerships among community organizations.

Qualities of partnerships

Three qualities are posited as being indicative of partnerships: closeness, equity, and integrity. The quality of closeness is conceptualized as ranging from "unaware of the other party" through "transformational" (see Figure 3.2). In relationship theory, closeness is a function of three components: frequency of interaction, diversity of interaction, and strength of influence on the other party's behavior, decisions, plans, and goals.[5] Although simply being aware of another person or entity (for example, a policeman or a government regulatory agency) can influence one's behavior (following the regulations, for example), the more typical case takes place when influence occurs through personal interaction. Frequency of interactions is an important but incomplete index of closeness. Parties who do many different types of activities together are closer than parties who interact just as frequently but always do the same activity. Thus, relationships are closer when they have a diverse basis of interacting with each other, that is, when they grow beyond the original focus (such as student placements in a service-learning class), identify additional projects and diverse activities on which to work, and develop a broader network of relationships for collaboration.

In addition to frequency and diversity of interactions, relationships that demonstrate interdependency, bilateral influence, and consensual decision making are even closer. These relationships are characterized by a transition from a tit-for-tat pattern of appraising personal outcomes according to one's own gains (a transactional basis) to a consideration of joint outcomes, a communal attitude, and accommodation that supports mutual trust and a long-term perspective.[6] The highest order for these interactions is when they result in the merging, growth, and transformation of the entities that share a common fate.[7]

The quality of equity, a second dimension of the relationship, raises the issue that the contributions and outcomes of interactions will be quantitatively and qualitatively different for each party, and the standards against which they are appraised will be unique for each party. Equity theory posits that even when the inputs and outcomes are unequal, when outcomes are perceived as proportionate to inputs and those ratios are similar, a relationship is satisfying. Helping interactions are inherently asymmetrical and inequitable in that someone with resources is often helping someone who lacks that resource;[8] thus, one party invests disproportionally more and one party receives disproportionally more. In general, transformational partnerships reflect equity in that both parties view the interactions as fair and demonstrate growth in ways that are uniquely meaningful to each. Thus, equity is a more reasonable and practical aspiration for civic engagement activities than is equality.[9] Equity highlights newer models of civic engagement that are not just working "to and for" communities, but rather are working "in and with" communities toward mutually satisfying goals and reciprocal interactions. Thus, campus-school relationships that are equitable are more egalitarian and reciprocal, with both parties giving and receiving.

A third quality of partnerships is integrity. Morton argues that relationships lack integrity when they are paternalistic, self-centered, produce negative consequences, create dependencies and false expectations, and leave others tired and cynical.[10] In contrast, relationships with high levels of integrity possess "deeply held, internally coherent values; match means and ends; describe a primary way of interpreting and relating to the world; offer a way of defining problems and solutions; and suggest a vision of what a transformed world might look like."[11]

The degree of integrity could vary across each of the six relationships identified in Figure 3.1. Although the types of relationships portrayed in Figure 3.2 can vary on integrity and equity, we posit that the closer the relationship is, the greater the integrity and equity, with transformational relationships always having high degrees of equity and integrity.

Prior to the opening of GWCHS, an evaluation of IUPUI activities in the Near Westside neighborhoods, based on interviews with key community leaders and residents, provided evidence that the relationship between the campus and community was growing closer during this period. Virtually all respondents recognized a greater level of interaction between the Near Westside and IUPUI as a result of HUD-COPC activities (mean = 4.62 on a 5-point response scale). One respondent described weekly, if not daily, contact between neighborhood leaders and IUPUI faculty and staff. Thus, frequency of interactions was higher. Almost all community respondents agreed that the Near Westside and IUPUI had jointly participated in multiple types of activities over the past year (4.32 out of 5.00). Some of the collaborative activities mentioned were tutoring programs, job and health fairs, grant writing, and community meetings. Thus, diversity of interactions had also increased. A more neutral position was taken in answer to inquiries about evidence of the Near Westside and IUPUI influencing each other's functioning (3.32 out of 5.00). In addition, some respondents felt that the community benefited more than the university did (that is, the interdependency was asymmetrical) at this stage of the relationship. Most respondents believed that IUPUI and the Near Westside had been able to discuss difficult issues with each other during the past three years (4.30 out of 5.00). There was also a sense that openness had increased over time and that meaningful progress had been made.

At that time, two themes emerged regarding ways IUPUI should work to improve its relationship with the Near Westside. The first was that the university needed to demonstrate commitment to the partnership with the community beyond the scope of past activities. The second was that the community needed to clarify its needs and consider whether the university was well suited to contribute to them. Some respondents observed that the community sometimes changed its position midway through a project, contributing to dissatisfaction with outcomes.

There has been growth in the relationship between IUPUI and the Near Westside, and much of this growth has been simulated by the opening of GWCHS. The success of the GWCHS-IUPUI partnership inspired GWCHS staff to develop new ways to engage youth in the community, including service-learning classes and service events, and the community in GWCHS. IUPUI has also helped secure funding for community programs in financial literacy and health promotion. The achievements are communicated to GWCHS parents and neighborhood residents through school press releases, local community newspapers, and reports at community meetings like the community advisory council and the WETF. The partnership and continued need to inform more people has also prompted ONP to develop an e-newsletter that informs the campus and community stakeholders about the work by the partnership.

The development of these interactions between IUPUI and GWCHS over time has reflected more frequent and more varied activities; however, it is probably the case that IUPUI has been more involved in the life of GWCHS than GWCHS has been in the life of IUPUI. Nevertheless, both share a genuine concern for educating youth and creating opportunities for the entire commu-

NEW DIRECTIONS FOR YOUTH DEVELOPMENT • DOI: 10.1002/yd

nity to be engaged in growth and development activities in ways that reflect mutually held values and goals. Most programs have been developed from conversations among the stakeholders. One community leader notes that the partnership works on the Near Westside because everyone is bringing and taking away something. She equates the partnership to a potluck dinner: each person brings one dish, and by the end of the meal, each walks away full. This illustration demonstrates the diversity of ideas as well as the commitment that partners bring to the partnership.

The impact of the collaborations between GWCHS and IUPUI has been significant. For example, 88 percent of the 2007 graduating class went on to postsecondary education, including some to IUPUI, exceeding local and national norms. Jim Grim, the community school coordinator and learning communities initiative director, said:

The Sam H. Jones Community Service Scholars from IUPUI serve as important role models in addition to the invaluable information they presented to our students about college and its quality-of-life benefits. We would not have had 80 percent of our first graduates in 2006 and 88 percent in 2007 go into postsecondary education had it not been for the ongoing presence of IUPUI service students at Washington. Our success is the direct result of IUPUI's commitment to service learning and we are grateful for it.

College students have been inspired by the partnership with the school as well. One IUPUI Fugate Scholar, a scholarship instituted by the university in 2006 to support student involvement at the school, said, "I am proud to speak about my entire George Washington experience. The fact that we are able to be positive role models is great just with our presence, but the fact that we're able to convey the message about furthering their education is the most important issue we're addressing."

This mutually beneficial partnership continues to have a positive impact on youth development and academic achievement. Ninety-one percent of the 2008 graduating class pursued postsecondary education. According to documentation provided to the Community Advisory Council, there has also been a noticeable increase in attendance and math and science standardized test scores.

Conclusion

In analyzing the dimensions of the GWCHS-IUPUI partnership, we have proposed a network that identifies four constituencies (campus, school, community organizations, residents) and six relationships among them. Furthermore, three dimensions are posited to reflect the quality of relationships that have developed into partnerships: closeness (a function of frequency of interactions, diversity of interactions, and interdependency), equity, and integrity. Evidence has been presented that bears on the quality of the six relationships in the case of GWCHS. Prior to the opening of GWCHS, the relationships among IUPUI, residents, and community organizations were developing with evidence of increasing and diverse interactions, and common purpose reflected in the importance of enhancing educational opportunities in the community through IUPUI's civic engagement in the Near Westside neighborhoods. Although all of these relationships might not have been symmetrical, they were appraised as beneficial and equitable. Furthermore they were developing qualities of high integrity. Residents; staff from community centers, public school administration, and other community organizations; and IUPUI representatives were working together in a concerted way to meet the challenge of having no schools in the neighborhoods, forging a common vision of opening schools, and developing strategies for working toward a solution. Thus, there was clear evidence that they were working with one another and with an integration of purpose. Furthermore, partnerships (not just relationships) were being established.

When GWCHS opened, the network of existing partnerships contributed to the development of the school as a community school. With strong school leadership and the active leadership of the WETF and representatives from IUPUI and the community, CWCHS was able to thrive within this well-functioning network of partnerships and build new working relationships between GWCHS and the other three sets of constituencies. The evidence that is presented supports the conclusion that the relationships between GWCHS and IUPUI, residents, and community organizations are close, reciprocal partnerships with integrity.

NEW DIRECTIONS FOR YOUTH DEVELOPMENT • DOI: 10.1002/yd

Several indicators point to how the high quality of these partnerships has contributed to the growth of the constituencies. GWCHS was awarded the Inaugural National Community School Award by the National Coalition for Community Schools in 2006 and was recognized by the KnowledgeWorks Foundation of Cincinnati, Ohio, in 2004 as "one of the nation's best examples of a school as center of community." In 2008, the U.S. Department of Education notified GWCHS partners that they were one of ten community schools, and the only one in the Midwest, to be awarded $2.4 million over the next five years in the nation's first federal full-service community schools funding authorized by Congress. A community organization, Mary Rigg Neighborhood Center, which employs the GWCHS community school coordinator, will serve as fiscal agent on behalf of the extensive collaborating partnerships. The federal funding expands support services for GWCHS students, families, and residents, particularly after school and on weekends. The grant also includes a five-year evaluation led by IUPUI's Center for Urban and Multicultural Education and recognizes GWCHS as a model for school-community partnerships that will be replicated at three additional community high schools in IPS.

In the near future, the partners hope to secure funding for a graduate student from the School of Social Work to assume an intermediary role as community school coordinator at Wendell Phillips Elementary School 63, thus replicating the GWCHS model. The Near Westside received a Great Indy Neighborhoods Initiative (GINI) grant to aid with resident-based community planning, including education. Finally, the students at GWCHS have shown growth that leads us to conclude that this model of engagement has an impact on the academic achievement and overall development of youth. In the past three years, the school has documented a 70 percent decline in the percentage of students required to see the on-site probation officer. Youth involvement has increased in community service, including the establishment of the Key Club, participation of over eighty youth in Make a Difference Day, procurement of a Youth as Resources grant by young people who adopted a nearby park, and participation of youth in the GINI quality-of-life planning and implementation.

The partnership with GWCHS has significantly contributed to the ability of IUPUI to be an engaged campus. Through continued communication between staff in the Center for Service and Learning and community leaders in the Near Westside, new programs are easily discussed and implemented, demonstrating the integrity of the partnership. As a priority partner, GWCHS is the host site for many new programs. This partnership has prompted IUPUI to become involved in larger national and global conversations on engaging universities in underresourced neighborhoods. The civic engagement work in communities, including work in the Near Westside and with GWCHS, has been the basis for numerous recognitions to the campus, among them the Presidential Award for Community Service, the Carnegie Foundation Classification for Community Engagement, the Saviors of Our City award, recognition in *Colleges with a Conscience*, and *US News and World Report* recognition for service-learning each year since 2002.

Although this analysis has focused on the four cardinal points and the six relationships among them in Figure 3.1, this is nevertheless a simplistic approach to understanding the complexity of university-school partnerships. First, many of the activities draw in representatives from more than two of the cardinal points, transcending the dyadic characterization. Second, each cardinal point can be elaborated. For example, GWCHS has had an influence on other schools in IPS, serving as a resource and a model for other schools to become community schools. Because of the activities in the neighborhood since GWCHS opened, there have been additional unique collaborations between and among community organizations as they have worked together on activities. Resident involvement in GWCHS has spawned involvement in other educational initiatives such as charter schools and strategic planning in health, safety, and other areas. The GWCHS Alumni Association is now active in bringing community support to the school, and neighborhood residents have a swimming pool, a wellness center, and a community meeting room available for free use. Finally, the GWCHS-IUPUI partnership has resulted in the formation of an IUPUI P-16 council to coordinate campus work with other schools in Indianapolis and

NEW DIRECTIONS FOR YOUTH DEVELOPMENT • DOI: 10.1002/yd

central Indiana. The GWCHS community school coordinator is an active member of this council. Thus, there has been growth in the individual missions of the major constituencies through the development of strong partnerships, and this growth has strengthened their roles with similar constituencies.

Jacoby differentiates between relationships that are merely transactional and those that are transformational.[12] Transactional relationships are instrumental in design, focused on accomplishing bounded tasks in a way that benefits everyone; transformative relationships "invite the possibility that . . . joint work" may well change individuals, relationships, and organizational contexts as new questions are considered, problems are redefined from new perspectives, identities and meanings are challenged and reconstructed, and new possibilities are envisioned. Transformational "partnerships have the ability not to just get things done but to transform individuals, organizations, institutions, and communities." Our analysis of the six relationships that have resulted from the GWCHS-IUPUI collaboration suggest that each of them has moved from relationship to partnership because they are close, equitable, and high in integrity. The analysis of this case study of GWCHS suggests that its work with IUPUI, residents, and community organizations has not only transcended the transactional but may indeed be a transformational network for change.

Notes

1. Kellogg Commission on the Future of State and Land-Grant Institutions. (1999). *Returning to our roots: The engaged institution.* Washington, DC: National Association of State Universities and Land-Grant Colleges.

2. Cruz, N., & Giles, D. E. (2000). Where's the community in service-learning research? *Michigan Journal of Community Service-Learning,* [Special issue] 2, 28–34.

3. Noam, G., & Tillinger, J.R. (2004). After-school as intermediary space: Theory and typology of partnerships. In After-School Worlds: Creating a New Social Space for Development and Learning. Gil G. Noam (ed.). *New Directions for Youth Development,* No. 101, 75–113.

4. Portland State University. (2008, March). A guide to reciprocal community-campus partnerships: Proceedings from Portland State University's Partnership Forum. Retrieved May 4, 2009, from http:// http://depts. washington.edu/ccph/pdf_files/Guide_corrected_041808.pdf.

5. Berscheid, E., Snyder, M., & Omoto, A. M. (1989). The Relationship Closeness Inventory: Assessing the closeness of interpersonal relationships. *Journal of Personality and Social Psychology, 57*, 792–807.

6. Clark, M. S., & Mills, J. (1979). Interpersonal attraction in exchange and communal relationships. *Journal of Personality and Social Psychology, 37*, 12–24; Rusbult, C. E., Verette, J., Whitney, G. A., Slovik, L. F., & Lipkus, I. (1991). Accommodation processes in close relationships: Theory and preliminary empirical evidence. *Journal of Personality and Social Psychology, 60*, 53–78.

7. Mashek, D., Cannaday, L., & Tangney, J. (2007). Inclusion of community in self scale: A single-item pictorial measure of community connectedness. *Journal of Community Psychology, 35*(2), 257–275.

8. Bringle, R. G., & Velo, P. M. (1998). Attributions about misery. In R. G. Bringle & D. K. Duffy (Eds.), *With service in mind: Concepts and models for service-learning in psychology* (pp. 51–67). Washington, DC: American Association for Higher Education.

9. Bringle, R. G., & Hatcher, J. A. (2002). University-community partnerships: The terms of engagement. *Journal of Social Issues, 58*, 503–516.

10. Morton, K. (1995). The irony of service: Charity, project, and social change in service-learning. *Michigan Journal of Community Service Learning, 2*, 19–32.

11. Muthiah, N. R., & Reeser, D. M. (2000). *IUPUI-WESCO COPC evaluation report.* Indianapolis, IN: IUPUI Center for Service and Learning.

12. Jacoby, B., & Associates. (Eds.). (1996). *Service-learning in higher education: Concepts and practices.* San Francisco: Jossey-Bass.

ROBERT G. BRINGLE *is director of the IUPUI Center for Service and Learning and Chancellor's Professor of Psychology and Philanthropic Studies.*

STARLA D. H. OFFICER *is a coordinator in the Office of Neighborhood Partnerships, IUPUI Center for Service and Learning.*

JIM GRIM *is the community school and learning initiatives coordinator at George Washington Community High School.*

JULIE A. HATCHER *is the associate director of the IUPUI Center for Service and Learning.*

The Agatston Urban Nutrition Initiative draws on the resources of a premier research university and the community to address the problem of obesity in children and youth. It takes a problem-solving approach to learning, in which university and community students work jointly in a curriculum that integrates in-class and service-learning to focus on problems that afflict society—in this case, obesity.

4

The Agatston Urban Nutrition Initiative: Working to reverse the obesity epidemic through academically based community service

Francis E. Johnston

THE HIGH LEVELS OF obesity in the United States are part of a global problem of truly enormous proportions, one that has increased dramatically over the past forty years and shows no sign of abating in this first decade of the new century. Whereas national surveys carried out in the 1960s by the federal government found some 15 percent of American adults to be overweight or obese, the 2003–2004 National Health and Nutrition Survey reported a prevalence of 66 percent—a fourfold increase—among those

NEW DIRECTIONS FOR YOUTH DEVELOPMENT, NO. 122, SUMMER 2009 © WILEY PERIODICALS, INC.
Published online in Wiley InterScience (www.interscience.wiley.com) • DOI: 10.1002/yd.306

twenty to seventy-four years old, of whom 33 percent were categorized as obese.

The increasing rates of overweight and obesity in the United States are similar to those found around the world. A joint 2002 report by the International Obesity Task Force and the European Association for the Study of Obesity noted that "obesity is rising at an alarming rate throughout Europe. It forms a pan-European epidemic that presents a major barrier to the prevention of chronic non-communicable diseases."[1]

Recently Kelly and coworkers estimated the prevalence of adult overweight and obesity in the world in 2005. Their analysis indicates that 33.0 percent of the world's adult population (1.3 billion people) was either overweight or obese. Although rates were higher in economically developed countries than among lesser-developed nations (overweight: 35 versus 20 percent, obese: 20 versus 7 percent), "the much larger population of developing countries results in a considerably larger absolute number of individuals affected." The authors say that if the trends continue, as many as 58 percent of the world's population—over 3 billion persons—could be overweight or obese by 2030.[2]

The burden of obesity

Obesity imposes a significant burden through its impact on the physical and psychological well-being of those affected and, more broadly, on society itself. Johnston and Harkavy have noted the following examples:[3]

- Obesity raises blood cholesterol and triglyceride levels, tends to raise the risk of cardiovascular disease, raises blood pressure levels, and can induce diabetes.
- Young adult obesity is associated with an increased risk of disability in the later years.
- Society discriminates against the obese in hiring, salaries, promotions, and employment. The overweight are seen as lacking

NEW DIRECTIONS FOR YOUTH DEVELOPMENT • DOI: 10.1002/yd

self-discipline, sloppy in their dress, disagreeable, and emotionally unstable.

- Medical expenses attributable to obesity accounted for 9.1 percent of total U.S. medical expenditures in 1998 and may have reached as high as $78.5 billion.
- Over 110,000 deaths from all causes were attributable to obesity in 2000.

Overweight and obesity in children and youth

Table 4.1 shows the percentage of overweight and obese American children and youth from the 1960s to 2003–2004. The approximately fourfold increase from the mid-1960s to the mid-1990s is clearly evident and is similar to the figures for the adult population.[4]

The health risks noted above are greater among adults since they are chronic diseases that appear and worsen with aging. However, such diseases are now being diagnosed more frequently among young people. In addition, overweight children and youth are subjected to at least the same degree of psychological abuse as are their adult counterparts. Given the likelihood that the results of this abuse will carry over to adulthood, the challenge is evident.

Finally, the accumulation of evidence over the past fifty years points clearly to the significant likelihood that obesity during adolescence will carry over into adulthood and that such early-onset

Table 4.1. Prevalence of overweight among children and adolescents, selected years

Age	1963–1965	1971–1974	1976–1980	1988–1994	1999–2000	2001–2002	2003–2004
2–5 years		5	5	7.2	10.3	10.6	13.9
6–11 years	4.2	4	6.5	11.3	15.1	16.3	18.8
12–19 years	4.6	6.1	5	10.5	14.8	16.7	17.4

Source: Retrieved May 11, 2009, from http:// www.cdc.gov/nchs/products/pubs/pubd/ hestats/overweight/overwght_child_03.htm#Table%201.

obesity is especially difficult to reverse. Do overweight and obese youth become overweight and obese adults? Over the years, a number of researchers have studied the probability that overweight and obesity during childhood persist into the adult years.[5]

Another approach to expressing the risk of adult obesity among obese children was presented in an earlier review of the literature by Serdula and his colleagues.[6] They concluded that:

- About a third of obese preschool children and about half of obese school-age children become obese as adults.
- The risk of adult obesity was at least twice as high for obese children as it was for those who were not obese.
- The later in childhood that obesity is present, the greater the likelihood is that obesity will continue in adulthood.

It is clear that obesity is more than just a health-related issue. It is a societal problem that has grown seemingly out of control to the extent that it may be characterized as the greatest public health failure of the twentieth century.

Obesity is more than a disease or a condition that shortens life expectancy, leads to discrimination and emotional distress, and imposes a major burden on the economy. It is also an indicator: a marker of a society whose structure intensifies and focuses its effects and a culture whose values create the conditions underlying the dramatic increases that are widely reported and described but not abated. And it is the obesity culture that is at the base of this complex and poorly understood problem which must be confronted through effective and sustainable efforts.[7]

Understanding obesity as a complex, ill-structured problem

Among those who deal with the formal process of problem solving, two general categories are recognized: those that are well defined (or well structured) and those that are ill defined (or ill structured).

The first category includes those that are fully, or almost fully, specified. Their major parameters (components) are established and may be organized into a structure that relates one to another in a logical scheme that leads to an outcome. But many of the problems that are faced in the day-to-day world fall into the ill-structured category. Although some, or even many, of the components of such problems may be known, they cannot be related to each other in any useful or predictive structure or model.

Ill-defined problems are known by a number of characteristics, including these:

- More information than is initially available is needed to understand the problem and decide what actions are required for resolution.
- No single formula exists for conducting an investigation to resolve the problem.
- As new information is obtained, the problem changes.
- One can never be sure that the right decision has been made.[8]

It is the complexity and lack of a clear structure that characterizes the problem of obesity and has stood in the way—and continues to do so—of developing and applying workable solutions. Rather than approaching obesity as a complex problem that resists a simple understanding and implementing strategies that are appropriate to its nature, it has been wrongly understood as well defined and linear—one that can be solved by exclusively applying models derived from reductionist, experimental (or quasi-experimental) research. Such models work well when the phenomena they explore can be specified and controlled and the hypotheses that direct the research are applicable to real-world situations. But they do little, if anything, to clarify the complexity of obesity and suggest workable solutions.

The failure of the many interventions based on an experimental model, clinical regimens, media investigations and reports, and commercially driven weight loss programs speaks volumes to the complexity of obesity as a problem to be solved. And the clustering of obesity and its accompanying diseases among the poor, the underserved, and the marginalized, particularly in developed economies, is a clear challenge to those who are concerned with health, welfare, and well-being. If decades of research have done nothing more, they have indicated clearly that the problem of overweight and obesity is not simply one of a person's diet, character failings, genes, or the manipulation of his or her behavior by the food industry. The problem of obesity is a broadly based cultural one, with causal agents identified throughout society that interact not as a simple chain, but through a complex network of beliefs, values, and behaviors that form the context responsible for the defining health issue of our time.

In "Culture and Economy in the Etiology of Obesity: Diet, Television and the Illusions of Personal Choice," Brown and Krick have written:

Cultural context includes many unconscious and taken-for-granted circumstances that greatly limit individual choice and behavior. For example, some things are simply considered "normal" and unquestioned parts of life including driving cars instead of walking, eating calorie-dense industrially produced foods, and watching television for hours per day. In the daily life of "normal" people within the cultural mainstream, an individual's daily choices primarily involve what car they drive or what route they take, which energy-dense food they eat, and what they will watch on television. The cultural and economic context—historically shaped by powerful socioeconomic forces [such as] corporations—constrains individual choices in habitual behaviors. At the same time, a consumer oriented capitalist economy establishes an illusion of personal choice about work, diet and activity patterns. It is difficult for individuals to swim against the current of cultural forces that lead toward fatness; it is a culturally constructed "obesogenic" environment.[9]

A comprehensive wellness program

The remainder of this article focuses on Agatston Urban Nutrition Initiative (AUNI), a comprehensive program of the University of Pennsylvania Netter Center for Community Partnerships to promote nutritional health and community well-being through partnerships with university-assisted community schools.

West Philadelphia: The population and its health

The design of any program whose goal is to improve health and well-being, no matter how comprehensive, how theoretical, or how elegant it may be, is place centered: it is implemented in a particular location, large or small. A place is far more than a set of geographical coordinates or a municipal entity. It is a system of interactions among physical, cultural, and demographic components. To different observers, such a place may be defined differently—for example, as the total environment, an ecosystem, a set of contextual variables, or a community whose members interact with each other at varied levels of intensity and in different roles. Regardless of the conceptualization, the design, implementation, and evaluation of any program requires an understanding of space and the interactions that exist. For that reason, in order to understand the AUNI, it is necessary to understand health-related aspects of West Philadelphia, the place where it was designed and is currently applied.

West Philadelphia is a fourteen-square-mile area west of Philadelphia City Hall and situated between the Schuylkill River and the western city limits. Its 220,000 residents (2000 census) live in twenty-five neighborhoods served by thirty-one public schools. The area is predominantly (non-Hispanic) black, with twenty-seven of its forty-eight census tracts recorded as greater than 90 percent black. Overall, the population is 72 percent non-Hispanic black, compared to 40 percent for the city as a whole.

The median U.S. national family income is more than twice that of West Philadelphia residents, and the national poverty rate is less than half. A similar picture is seen when West Philadelphia is compared to the city as a whole. West Philadelphia has a higher rate of poverty, a lower median family income, and a higher unemployment rate than Philadelphia as a whole, along with almost twice the percentage of vacant residential properties of the city.

Insofar as health is concerned, racial/ethnic health disparities in Philadelphia are similar to those observed in other major American cities. Low birth weight rates among black women in 2000 were almost twice the white rate (13.8 per 1,000 live births versus 7.1), and births to teenagers were two-and-a-half times the white rate (21.1 versus 8.6). Death rates among African Americans were higher than those of whites, by gender, for cardiovascular disease, stroke, diabetes, and HIV/AIDS.

One of the most useful bodies of data on the health of West Philadelphia comes from research conducted jointly by University of Pennsylvania and West Philadelphia school students as part of their participation in the AUNI. Especially in the early days of the initiative, students carried out formative research on the nutritional status of West Philadelphia middle and high school students to help determine the needs of the community and shape the structure of the program.

Taken jointly, this formative research presents a full picture of the nutrition-related problems of West Philadelphia black teenagers. Among eleven to fourteen year olds, the prevalence of obesity is significantly greater than is reported for the United States for all races combined, as well as that recorded among national probability samples of their African American age peers. Furthermore, from the 1970s to the 1990s, there was a threefold increase in the prevalence of obesity of males between eleven and fourteen years old and a fourfold increase in females of the same age.[10]

Analysis of the dietary intakes of middle school students has shown that the consumption of total fat, saturated fat, protein, and carbohydrate was not only higher than the recommended daily allowances but was also higher than that of their age peers nationally. And finally, deficiencies were found in the intakes of calcium, zinc, fiber, and vitamins A and D, nutrients that are important in promoting health and reducing the risk among females of osteoporosis, especially after menopause.[11]

The Agatston Urban Nutrition Initiative and the obesity culture

It is within this setting—this place—that AUNI was developed as a community-based program that has brought the resources of the University of Pennsylvania into a partnership with the West

Philadelphia community, in particular its schools. A number of core principles embodied with AUNI define its organization and its activities. First, AUNI operates in community schools, described by the Coalition for Community Schools "as both a place and a set of partnerships . . . with an integrated focus on academics, health and social services, youth and community development and community engagement. . . . Schools become centers of the community and are open to everyone—all day, every day, evenings and weekends."[12] Furthermore, AUNI operates in university-assisted community schools (UACS), in which the resources of the university help to educate, engage, empower, and serve all members of the community in which the school is located. The UACS partnership is a continuum of experience, from preschool through secondary school and the undergraduate and postgraduate years. University-assisted community schools provide a powerful means for universities to advance teaching, research, learning, and service, along with the civic development of their students.

Second, AUNI is based on academically based community service (ABCS), an approach that integrates the three major missions of the university—academics, research, service—into a single holistic framework and extends it to the community's schools throughout the range of grades by means of active learning and peer education (Figure 4.1). This approach gains further strength from an emphasis on problem-solving learning, a learning method in which students learn not just

Figure 4.1. Schematic of academically based community service

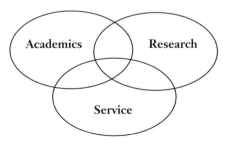

from memorizing and repeating facts but by a focus on working in a collaborative and democratic process to solve real-world problems.

Finally, as part of a problem-solving curriculum, students engage in service-learning activities that are tied directly into their course work, a relationship that transforms their classroom activities into a real-world context. In all cases, the school students are the agents of change in their families and the community as a whole.

AUNI in action. The following is an example of AUNI in action.

It is 7:00 am in West Philadelphia and the traffic from the western part of the city and the suburbs is beginning to increase notably as automobiles, buses and, below ground, the Market/Frankford subway line move toward the city center. Little activity is visible on the campus of the University of Pennsylvania, three blocks to the south of University City High School, and even less around UCHS and the adjacent Charles Drew School. The high school is the educational home of some 2,500 9–12-grade students, 93% of whom are African-American, 6% Asian, and the remainder Hispanic, White, and Native American. Of the families whose offspring attend UCHS, almost 90% are categorized as low income. Some 650 pupils from Kindergarten through the 8th grade attend Drew School with its ethnic and socioeconomic profile the same as those of University City High.

Twenty years ago, a barren lot separated UCHS and Drew, with a few patches of worn and neglected grass but mostly just dirt, unsuitable for anything. In 1998, two Penn recent alumni, Danny Gerber, a former Philadelphia elementary school teacher, and Tamara Dubowitz, a graduate student in anthropology at Penn, began a process that has transformed the lot into an urban school garden that now is the centerpiece of the Agatston Urban Nutrition Initiative. Dubowitz has since earned her doctorate in public health while Danny Gerber has stayed on to spearhead AUNI as the prime nutrition/lifestyle program in West Philadelphia.

As the clock moves beyond 7:00 am and approaches 7:30, even though there is still little activity at UCHS and Drew, one can observe students along with AUNI staff beginning to converge on the garden that occupies the space between the two schools. The students are from the high school and the university, with others who are undergraduate Fellows from institutions across the country. All are there to begin to work in the garden in activities that are appropriate for the season. They may work the soil itself, weeding the garden or turning over the soil in the raised beds. Or they may harvest the wide variety of vegetables, herbs, and spices that are grown, preparing them for sale by elementary school students from Drew in student-operated afterschool markets, or by street-corner vegetable stands in the community and on the Penn campus. And if it isn't the growing season, students will work indoors, growing the seedlings to be used in the spring or conducting curriculum-based hands-on learning activities such as cooking, learning to read and interpret food labels, and acting as "food soldiers," spreading messages about food and nutrition to students and staff throughout the school. In all cases they will work collaboratively with teachers and university students to solve the problem of nutritional well-being in modern, urban society as that problem affects them and their communities. The garden at University City High School is the largest and the most developed of those that are part of the Agatston Urban Nutrition Initiative. But it is only one of over ten that are part of AUNI, found across the city but primarily in Philadelphia. The gardens are linked to the various nutrition and health classes and, in almost every case, to an afterschool fruit and vegetable stand that is staffed and maintained by school students with high school and university students serving as peer teachers and facilitators.[13]

Figure 4.2 shows students working with harvested vegetables as part of AUNI.

Figure 4.2. High school students working with harvested vegetables as part of the Agatston Initiative

The AUNI began in 1990 as a course in anthropology, with eighteen undergraduates teaching nutrition and health to some forty sixth-grade students at a middle school not far from the Penn campus. It has since extended its activities to over twenty selected schools in West Philadelphia and elsewhere in the city; it is currently adapting and replicating its programming to cities elsewhere in the United States. Even as this expansion and transformation have taken place, its goals have remained essentially the same:

• To help students from kindergarten through the postgraduate years to enhance their nutritional well-being and develop healthy lifestyles
• To improve their educational experiences and increase learning by means of a curriculum that involves active, hands-on, service-centered learning
• To instill a greater civic awareness and participation in society through working to change the problems that afflict modern urban society
• To enhance specialized training in academic disciplines and research methods through a curriculum that is rooted in problem-solving learning
• To develop models for change that can be extrapolated, with appropriate modifications, to other settings nationally and internationally

AUNI activities, 2008. AUNI is a fluid and ongoing program, and though its goals have remained the same since its inception, new activities have been introduced and others that did not work out have been discontinued. The best way to look at its most recent activities is to download the 2007–2008 report of the Netter Center for Community Partnerships (http://www.upenn.edu/ccp/index.php).

Currently, AUNI's major activities fit into four general categories:

• *Integrated school day nutrition education and academically based community service.* Through Eat Right Now, the school district of Philadelphia's comprehensive nutrition education program, AUNI conducts programs in twenty Philadelphia public schools. The primary focus of Eat.Right.Now is increasing the nutrition knowledge of K–12 students. AUNI incorporates as many hands-on components (such as monthly healthy food tastings) into this program as possible. In many cases, university students enrolled in nutrition-related service-learning courses work with public school students to explore and address nutrition-related issues in the community.

- *Increasing access to healthy foods.* AUNI engages young people in organizing nutritionally better choices for their communities through school and community-based efforts. Through AUNI, public school students work to improve lunchroom choices and operate after-school fruit stands. AUNI also works with public school students to help neighborhood food stores create convenient healthy food stations and to operate community farmers' markets.
- *Increasing opportunities for participation in regular physical activity.* Through school day, after-school, and summer programs, AUNI improves opportunities for youth and families to exercise regularly. AUNI works with physical education teachers and school coordinators to improve exercise opportunities during physical education class and recess time. And through the community schools program of the Netter Center for Community Partnerships, AUNI offers family-oriented exercise classes during evening programs.
- *Youth-led organizing, peer education, and internships.* Increasingly people recognize the important role that youth can play as organizers of solutions to societal problems on a variety of levels: as the deliverers of social and educational services, the developers of model programs, and key informants to policy-makers. In addition to school day peer education, AUNI coordinates job training and youth leadership programs for high school students. The AUNI internship program engages teens in organizing better food choices in their communities by working after school for ten hours each week. AUNI interns combine direct service approaches, which include teaching classes in healthy cooking and growing healthy foods in school gardens for sale at farmers' markets, with advocating for

broader systems change. In spring 2008, AUNI high school interns organized the Youth Action Council for the Philadelphia Urban Food and Fitness Alliance. They have also been highly involved in youth organizing on a regional and national level.

Evaluation of AUNI. The evaluation of AUNI, as with any other comparable initiative, involves the systematic collection and analysis of data in order to answer two broad questions. The first is about the process: the organization, the dynamics, and the operation of the program. The focus here is on how the program outcomes and impacts were produced as a way to help interpret the results. The process component of an evaluation is developmental, continuous, and flexible, with a significant inductive component. In an ongoing program, it is used to effect modifications along the way. This is called a process evaluation.

The second broad question is about the impact of the program. Did it accomplish its goals? Answering this question allows one to judge the value of the program at its end or at some other designated point. This is usually called impact evaluation.

Of course, no program that seeks to bring about significant change in, say, nutritional status can—or should—be judged on the answers to just one of these questions. Answering the impact question tells the evaluator just what was accomplished and informs those who make the decisions whether it should be continued. Answering the process question tells the evaluator why the goals were or were not accomplished and informs decision makers as to modifications that might be made in the future.

Evaluating AUNI is not an easy task for a number of reasons. It was designed to solve a complex, ill-defined problem using a variety of initiatives, most of them implemented at different times. Consequently it does not fall within the scope of an experimental— or quasi-experimental—research project. Rather it is a dynamic program that has evolved over the more than fifteen years of its existence, with modifications made as internal and external conditions have changed. AUNI is action oriented, community based, and participatory, and outcome expectations need to be assessed from those perspectives.[14]

A further stricture on evaluation has been imposed from time to time, depending on funding sources. For a number of years, AUNI has been part of Eat.Right.Now, a citywide school-based effort to improve dietary intakes and the nutritional well-being of Philadelphia school children and families eligible for food stamps.[15] The evaluation is conducted to determine change in behavior and knowledge of students receiving nutrition education through our program. Evaluation of Eat.Right.Now, which reaches more than 275 public schools, is being conducted by an outside agency, and is comprehensive in nature, of necessity requiring an across-the-board common protocol. Although this does not prevent any other evaluation of AUNI, it does constrain significantly the scope of projects that can be undertaken in terms of school sites and methods. Nonetheless, a number of approaches have been used to provide data on the process and investigate its impacts.

One approach to assessing the success of a program is its recognition by external bodies and groups. AUNI has been nominated for recognition by public agencies as well as private foundations. In each case, external committees conducted site visits to the Penn campus and AUNI schools, with committees interviewing staff, teachers, students, and senior administrators. The results have included the following awards and citations:

- AUNI was cited by the Robert Wood Johnson Foundation in 2003 as one of four "promising models" for improving health and nutrition among children in the United States.
- The university-assisted school programming of the Netter Center, of which AUNI is a core component, was named in 2003 by the U.S. National Academy of Science as the winner of the inaugural W. T. Grant Foundation Youth Development Award.
- The Pennsylvania State Horticultural Society recognized the Agatston Urban Nutrition Initiative school garden at University City High School and Drew K–8 school in 2003 as the best school garden in Philadelphia.

NEW DIRECTIONS FOR YOUTH DEVELOPMENT • DOI: 10.1002/yd

- AUNI was cited by the Community Outreach Partnership Centers Program of the U.S. Department of Housing and Urban Development in 2004 for its collaboration with the local community, helping to improve it through empowering residents and the organizations that serve them and strengthening relationships between campus and community.
- AUNI was recognized by Campus Compact in 2005 as one of eight exemplary Campus Community Partnerships in the United States.

In addition, of the more than twenty-three universities and colleges nationally that have adapted and replicated the Netter Center for Community Partnerships model of university-assisted community schools for their own institutions, most have included a nutrition component developed along the lines of AUNI.

One ongoing process-oriented assessment of AUNI, carried out regularly since 2005, uses questionnaires and focus groups for teachers from partner schools. Following are excerpts from the 2006 groups:

"I'd like to comment on the nutrition lessons also. They [the AUNI staff] are both excellent educators and their lessons plans were outstanding because when they went into the classrooms they had hands-on activities, they had appropriate materials for kids to read and respond to, and the children learned a wealth of information . . ."

"The social studies aspect was wonderful. Some of the fruits and vegetables, they [the students] actually traced the origin and how it arrived here in America. And that was great because some of it tied right in with my social studies . . ."

"It's great to have kids cooking, and you can measure that success, because they're hanging out in here eating on their lunch break and after-school. Kids definitely seem to enjoy it. . . . I think it's really important to establish a cooking culture. They've been so isolated. . . . So it's really great that they're so interested."

Various analyses of the impact of AUNI on dietary intake and nutrition and health-related behavior have been designed and carried out by University of Pennsylvania students as part of their ABCS course work. In one study, two hundred twelfth-grade high school students who had been part of AUNI for four years completed twenty-four-hour recalls of their diets with the nutrient intakes calculated by appropriate software. The nutrient intakes of these high school seniors were healthier than were those of grade 9 students who were in their first year of UNI. For example, twelfth-grade students consumed twice as many daily servings of fruits and vegetables than did the ninth graders.

In another study, six- and seven-year-old first graders consumed up to three times the servings of fruit after school the day after the school store—which sold a range of in-season fruits—was open for purchases compared to other school days or to students from a non-AUNI control school.

One of the principles underlying AUNI is that participation leads to behavioral change associated with improved nutrition and overall healthier lifestyles. One early study examined the impact of participation in the design and construction of a school fruit stand on subsequent visits to the stand. Two classes of sixth graders, typically eleven and twelve year olds, planned and implemented the initial fruit stand at the Turner School in 1995. The number of their visits and purchases in the weeks following were compared to those of two other classes that had not taken part in the planning. All students had been exposed to schoolwide announcements of the store and its times of opening. Those who were participants were significantly more likely to visit the store in the following several weeks and make purchases. Participation in the AUNI has been shown to be associated with more positive attitudes and beliefs in themselves and a greater willingness to try new foods than was found in nonparticipants. And finally, the parents of first and second graders were more likely to know about and make purchases from the fruit stand if their children were involved in it. They were significantly more likely to learn about the stand from their children than from seeing it themselves.

Conclusion

The AUNI was developed to cope with the growing problem of obesity in the United States, especially among children. In its earliest stages, it was seen as an ABCS program that emphasized innovative approaches to nutrition education with a reduction in body mass index as the goal. Within a few years, participants began to understand that the problem was far more than one of calories and realized that basic changes in lifestyle would be required to meet the challenges represented by obesity, which became conceptualized as the tip of the iceberg. This has led to the knowledge that obesity, as one component of nutrition-related illness and disease, is a response to a culture in which body weight becomes a commodity, the built environment contributes to the problem, and racial/ethnic health and health care disparities further intensify the situation. AUNI is not a research program or simply an intervention. It is a program rooted in principles of engagement, participation, partnership, and education. It recognizes the need for a broad-based set of initiatives to change the structures seen in the prevalence of obesity.

Obesity is the greatest public health failure of the twentieth century. The Agatston Urban Nutrition Initiative seeks to reverse that failure by attacking the cause, not the result.

Notes

1. International Obesity Task Force and the European Association for the Study of Obesity 2002. *Obesity in Europe, the case for action.* Retrieved October 12, 2006, from www.iotf.org/media/euobesity.pdf.

2. Kelly, T., Yang, W., Chen, C.-S., Reynolds, K., & He, J. (2008). Global burden of obesity in 2005 and projections to 2030. *International Journal of Obesity, 32,* 1431–1437.

3. Johnston, F. E., & Harkavy, I. (2009). *The obesity culture: Strategies for change—Public health and university-community partnerships.* London: Smith-Gordon.

4. U.S. Centers for Disease Control. Retrieved October 12, 2006, from http://www.cdc.gov/nchs/products/pubs/pubd/hestats/overweight/overwght_child_03.htm.

5. For example, see Must, A., & Colclough-Douglas, S. S. (2001). Adult health sequelae of pediatric obesity. In F. E. Johnston & G. D. Foster (Eds.), *Obesity, growth and development* (pp. 185–198). London: Smith-Gordon.

6. Serdula, M. K., Ivery, D., Coates, R. J., Freedman, D. S., Williamson, D. F., & Byers, T. (1993). Do obese children become obese adults? A review of the literature. *American Journal of Preventive Medicine, 22,* 167–177.

7. Johnston & Harkavy. (2009).

8. Gallagher, S. A. (1997). Problem-based learning: Where did it come from, what does it do, and where is it going? *Journal for the Education of the Gifted, 20,* 332–362.

9. Brown, P. J., & Krick, S. V. (2001). Culture and economy in the etiology of obesity: Diet, television and the illusions of personal choice. In F. E. Johnston & G. D. Foster (Eds.), *Obesity, growth and development* (p. 111). London: Smith-Gordon.

10. Gordon-Larsen, P., Zemel, B. S., & Johnston, F. E. (1997). Secular change in stature, weight, fatness, overweight, and obesity in urban African-American adolescents from the mid-1950s to the mid-1990s. *American Journal of Human Biology, 9,* 675–688.

11. Johnston, F. E., & Hallock, R. J. (1994). Physical growth, nutritional status, and dietary intakes of African-American middle school students from Philadelphia. *American Journal of Human Biology, 6,* 741–748.

12. Coalition for Community Schools. *Frequently asked questions about community schools.* Retrieved February 12, 2009, from http://www.community schools.org/index.php?option=content&task=view&id=6&Itemid=27#WhatCS.

13. Adapted from Johnston & Harkavy. (2009).

14. For an excellent discussion of issues in evaluating community-based programs, see Mittelmark, M. B., Hunt, M. K., Heath, G. W., & Schmid, T. L. (1993). Realistic outcomes: Lessons from community-based research and demonstration programs for the prevention of cardiovascular diseases. *Journal of Public Health Policy, 14,* 437–462.

15. Summary of *Eat.Right.Now.* Retrieved February 16, 2009, from http://www.actionforhealthykids.org/resources_profile.php?id=547.

FRANCIS E. JOHNSTON *is professor emeritus of anthropology at the University of Pennsylvania and distinguished senior fellow at the university's Barbara and Edward Netter Center for Community Partnerships.*

When Dayton Public Schools committed to return to neighborhood K–8 schools, the community organized to refocus many youth programs in schools and neighborhoods.

5

Dayton's Neighborhood School Centers

Dick Ferguson

IN 2004, the Fitz Center for Leadership in Community at the University of Dayton (UD) was awarded a $250,000 grant over two years (2004–2006) from the Dayton Foundation and asked to lead planning for Dayton's Neighborhood School Centers project. In 2006, a three-year pilot ending in 2009 was initiated with a budget of $1.2 million using the community school concept advocated by the Coalition for Community Schools. The Dayton Foundation, Dayton Public Schools, City of Dayton, Montgomery County, and sixteen foundation and corporate supporters are partners with the Fitz Center in a bold initiative to reconnect five Dayton public elementary schools to their neighborhoods after more than thirty years of court-ordered busing and to create full-service, year-round opportunities for neighborhood families and youth at these new schools. New school buildings, funded by a local levy and tobacco lawsuit settlement funds awarded through the Ohio School Facilities Commission, are scheduled to open through 2010, but the programming began in fall of 2006 (see Figure 5.1).

NEW DIRECTIONS FOR YOUTH DEVELOPMENT, NO. 122, SUMMER 2009 © WILEY PERIODICALS, INC.
Published online in Wiley InterScience (www.interscience.wiley.com) • DOI: 10.1002/yd.307

Figure 5.1. Overview of Dayton's Neighborhood School
Centers project

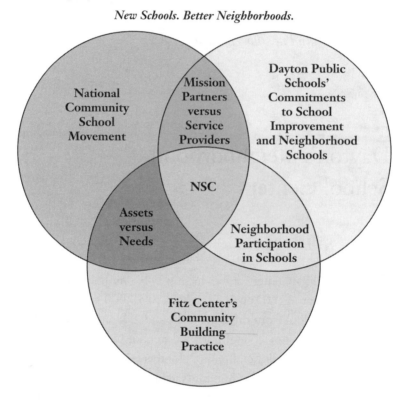

New Schools. Better Neighborhoods.

A community's project

The shared community vision is simple: New public schools are the
centers of their Dayton neighborhoods, serving as healthy places
of learning for children and families. The Neighborhood School
Centers mission is more complex: *Committed to children and families, we work with many partners to develop inventive, enduring relationships to create environments where students will excel and
neighborhoods will flourish with schools as their centers.*

The three-year pilot has these objective:

1. Secure start-up funding
2. Achieve strong involvement
3. Identify and remove policy barriers
4. Identify and leverage neighborhood assets
5. Plan and open new schools, and
6. Align with academic outcomes

Over the next five years, the objectives are:

1. Improve quality of life in the neighborhoods.
2. Attract families with school-aged children.
3. Improve student performance.
4. Realign community resources to support youth achievement.
5. Sustain leadership and support for Neighborhood School Centers.
6. Serve as a replicable national model.

In order to establish some intermediate measures of success, the Neighborhood School Centers focused on three outcomes: young people succeeding, neighborhood schools as the first choice of residents for their children's schooling, and neighborhood schools as centers of community involvement.

The Neighborhood School Centers operational plan was developed to sustain the relationships critical to successful outcomes. It is truly the community's project, with partners playing these various guiding and managing roles (see Figure 5.2):

Oversight Council. Recruited and organized by the Dayton Foundation, these individuals represent a cross-section of community leaders responsible for guiding the initial organization of the Neighborhood School Centers and maintaining a community commitment to the shared vision and objectives.

NEW DIRECTIONS FOR YOUTH DEVELOPMENT • DOI: 10.1002/yd

Figure 5.2. Partners for Dayton's Neighborhood School Centers project

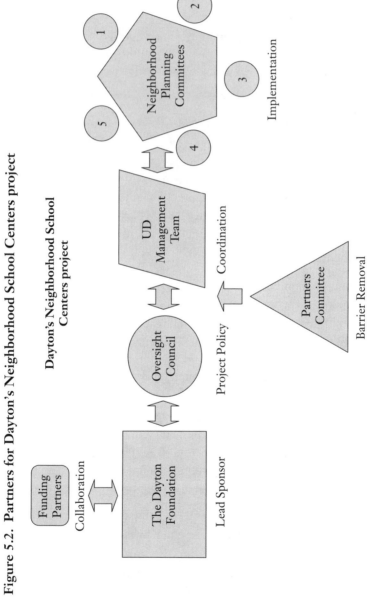

Dayton's Neighborhood School Centers project

Neighborhood Planning Committees. Recruited and organized by the Fitz Center, with appointments by Dayton Public Schools, neighborhood associations, and lead agency partners, these committees guide the work of the five sites and prioritize the efforts of the site coordinators. Where possible, they also serve as the building planning committees to advise Dayton Public Schools and its contract architects on the location and design of the new school buildings. Parents, teachers, and principals serve on these committees.

Partners Committee. Key public and private partners meet monthly with the superintendent of Dayton Public Schools to work out the administrative details of the Neighborhood School Centers. The partners are represented by their chief executive officers or senior staff delegates. This group's primary mission is to remove policy barriers.

UD Management Team. The Fitz Center for Leadership in Community at UD works on behalf of the partners to administer the organization and implementation of the community's vision for Neighborhood School Centers. The management team includes veteran community leaders who administer the pilot initiative, a respected social worker trained in consensus organizing, and a university professor with extensive public school experience.

Community building precedes programming

The creation of effective community partnerships that operate democratically and are highly inclusive is a daunting challenge. It is our opinion that most community collaborations fail because they do not pay adequate attention to building and sustaining relationships. Strong community partnerships resemble good friendships. Widely shared visions supported by citizens and professionals who enjoy working together are essential. To achieve a shared vision, citizens and professionals alike usually have to relinquish some aspect of their own beliefs and practices.

There is a not-so-obvious reason that the Fitz Center for Leadership in Community was asked to lead this effort. It is not because we are experts, advocates, or even thoughtful critics of public urban education. It is because we approach all of our work as a challenge of community building. The Fitz Center defines its mission as "community building in urban Dayton for the purposes of increasing the community's capacity for change and providing a rich context for experiential learning for students and faculty." Consensus organizing techniques are used in a variety of community settings. Dozens of Fitz Center projects are in progress in Dayton at any given time. Each emphasizes one or more of the five community building blocks identified through staff experience over the past three decades. These practices have become guiding principles of the Neighborhood School Centers. Each is described in detail in books by the authors noted below:

- *Developing community assets.* John McKnight and John Kretzmann of Northwestern University's Center for Urban Affairs and Policy Research initiated a national movement to focus on a community's assets instead of its needs when developing devastated communities. The same thinking informed developmental assets for youth.[1]
- *Strengthening social capital.* Harvard's Robert Putnam documented the loss of social capital in the United States in his book *Bowling Alone: The Collapse and Revival of American Community.* Trust, information sharing, reciprocity, and some shared norms characterize this illusive form of capital that Putnam says is essential to strong communities.[2]
- *Balancing inquiry and advocacy.* In his famous book, *The Fifth Discipline: The Art and Practice of the Learning Organization,* Peter Senge describes conversational skills that build relationships and improve understanding. Good questions, it seems, may be as important as good arguments.[3]
- *Cultivating leadership for adaptation.* Reviving communities means that people and institutions must change. The style of leadership that helps communities hold on to what is precious and let

NEW DIRECTIONS FOR YOUTH DEVELOPMENT • DOI: 10.1002/yd

go of the nonessential is described by Harvard's Ronald Heifetz in *Leadership Without Easy Answers* as adaptive.[4]

• *Finding a shared vision based on mutual self-interest.* Community organizing, long associated with the conflict organizing techniques of Saul Alinsky, has a new style described by its champion, Michael Eichler, in *Consensus Organizing: Building Communities of Mutual Self Interest.* This style works toward a shared vision as opposed to the vision of a dominant group or individual.[5]

We have taken the best insights of these creative thinkers and experimented with them in our work. The process requires new thinking on the part of experienced community leaders and constant practice by the citizens and professionals doing the work at the neighborhood sites. Compromise is essential. The Fitz Center uses the skills of consensus organizing to build community. Consensus organizers identify mutual self-interest to build and sustain community work. Community building, for the Fitz Center, is the art of co-creating a desired community future based on a widely shared vision. We use consensus organizing to get to the shared vision.[6]

Recognizing the opportunities

Dayton's Neighborhood School Centers are ideal for the work of community building. The opportunity to adapt the community school model in Dayton arose from the reality that Dayton neighborhoods are highly segregated racially. Court-ordered busing to achieve racial integration of schools was in place for more than thirty years. The order was lifted in 2002 due, in part, to the fact that the Dayton Public School system had become predominantly African American and leaders and the federal district court saw nothing further to be gained from mandatory busing.

The board of education and Dayton Public Schools leadership offered a return to neighborhood schools in exchange for passing a school building levy. If approved, funds from this capital levy

NEW DIRECTIONS FOR YOUTH DEVELOPMENT • DOI: 10.1002/yd

would be matched by the State of Ohio and enable the construction of new buildings throughout the district. The levy passed, but there was no plan in place to deliver on the campaign promise that schools would once again be centers of neighborhoods and would be available for community use. Dayton Public Schools appealed to the Dayton Foundation for help. The foundation turned to the Fitz Center because of its capacity for community building in urban neighborhoods. The center staff members have many years of collective experience in Dayton as organizers and facilitators.

A key challenge at the outset was to distinguish the effort to reconnect schools and neighborhoods from school reform. Although the academic achievement of Dayton Public School students was, and still is, below local expectations and statewide standards, the Neighborhood School Centers could not accept responsibility for improving standardized test scores. Although effective after-school programs, improved early childhood education, healthier students, and safer neighborhoods could be expected to help student achievement, project leaders distinguished the initiative from school reform efforts such as the creation of charter schools, which are popular and numerous in our urban community.

The Fitz Center accepted the leadership role while insisting the project become something more than the latest school reform initiative—the acknowledged agenda of most of the sponsors and community leaders. Experience in developing a partnership with Patterson-Kennedy Elementary School had taught many at UD the importance of defining the university's role as partner versus expert advisor or reformer. Teachers, administrators, and parents of elementary school students had experienced wave upon wave of reform efforts and had grown both skeptical of and resistant to change proposed by community leaders, foundations, universities, and other outsiders. Clearly any significant change had to come from within the Dayton Public Schools with the support of the community. The school levy promise to connect the public schools to neighborhoods and rebuild schools was an opportunity to build bonding social capital within schools and neighborhoods and develop bridging social capital with hundreds of agencies and associations committed to urban children, their families, and their neighborhoods.[7]

Selecting schools and neighborhoods

As a community, the Neighborhood School Centers project leaders selected the five schools and the targeted neighborhoods. Dayton Public Schools and the Dayton community needed a winning idea. The pilot schools and neighborhoods had to be immediately recognized as neighborhoods and schools that would eventually succeed as neighborhood school centers. It was no time for long shots. We developed these selection criteria:

- K–8 school site (not high school because no Dayton high school is neighborhood centered)
- Current institutional partners
- Other improvement efforts in the neighborhood
- Geographical spread of schools and neighborhoods
- Active neighborhood associations
- New building yet to be planned, if possible

Dayton boasts fifty-five well-organized neighborhoods, but not all of these can be redeveloped in the short run. The challenges of poverty, unemployment, foreclosures, and homelessness have hit Dayton hard. Nevertheless, we could readily identify a few Dayton neighborhoods that had major redevelopment potential and, in some cases, comprehensive community development under way. Since it was vital to the city and the school system to demonstrate successful connection of neighborhoods to new schools, we opted to focus on strength and select neighborhoods with significant community assets and short-term redevelopment potential. Neighborhood redevelopment was a key opportunity to link schools and neighborhoods.

NEW DIRECTIONS FOR YOUTH DEVELOPMENT • DOI: 10.1002/yd

Finding nonprofit program partners

Program opportunities were also available. Most youth-focused public and nonprofit agencies in Dayton worked in the public schools only under systemwide contracts. Funding for contract services other than those that had a direct impact on student academic performance had largely disappeared from the landscape of the elementary schools. Recreation, arts, after-school programs, clubs, and other extracurricular activities were victims of budget cuts in previous years. Yet many agencies and associations indicated an interest in locating services and programs in schools. Could targeted partnerships be developed that linked schools and agencies based on mutual self-interest? Would an agency offer a program at a neighborhood school without grant support, recognizing that unparalleled access to children and their families in the neighborhood setting was possible if the school could serve as the program site? Could the school building become a center of neighborhood life as it had been in the days before cross-district busing? The answer to all of these questions was yes (see Figure 5.3).

Many key community leaders were willing and able to broker a new community school model. The Coalition for Community Schools offered examples of success in other communities. The realization that Dayton had an opportunity to reposition schools as centers of neighborhood life convinced leaders to support the Neighborhood School Centers plan. Neighborhood association leaders, foundations, hospitals, universities, business leaders, and public officials added their support during the early phases of the organizing effort. A shared vision and multiple cases of mutual self-interest enabled a partnership approach to be taken.

Key partnerships

While there are many opportunities for collaboration in a community school, well-articulated partnerships are at the core of the most important community relationships. In Dayton's Neighborhood

NEW DIRECTIONS FOR YOUTH DEVELOPMENT • DOI: 10.1002/yd

Figure 5.3. Dayton's Neighborhood School Centers

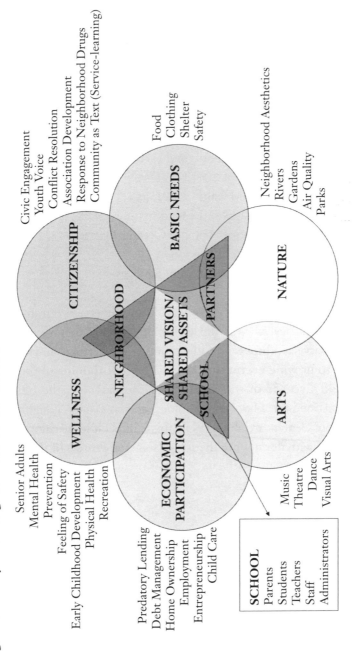

CITIZENSHIP
Civic Engagement
Youth Voice
Conflict Resolution
Association Development
Response to Neighborhood Drugs
Community as Text (Service-learning)

BASIC NEEDS
Food
Clothing
Shelter
Safety

NATURE
Neighborhood Aesthetics
Rivers
Gardens
Air Quality
Parks

WELLNESS
Senior Adults
Mental Health
Prevention
Feeling of Safety
Early Childhood Development
Physical Health
Recreation

ECONOMIC PARTICIPATION
Predatory Lending
Debt Management
Home Ownership
Employment
Entrepreneurship
Child Care

ARTS
Music
Theatre
Dance
Visual Arts

NEIGHBORHOOD

SHARED VISION/SHARED ASSETS

PARTNERS

SCHOOL

SCHOOL
Parents
Students
Teachers
Staff
Administrators

School Centers, partnerships between public schools and nonprofit partners are essential.

The key partnership at each site is between a lead nonprofit and the elementary school. This partnership provides for the selection, employment, supervision, and evaluation of the site coordinator. Each agency partner leads only one Neighborhood School Centers initiative. By selecting an agency trusted and valued in each specific neighborhood and school, the project was able to establish itself within the neighborhoods and schools within the first year.

These are the five schools and their agency partners:

- Fairview: Unified Health Solutions
- Edison: Dayton Urban League
- Kiser: The Salvation Army of Greater Dayton
- Cleveland: YMCA of Dayton
- Ruskin: East End Community Services

The partners bring considerable and diverse programming to their sites; nevertheless, the leadership team makes it clear that each is to serve as a broker for other community partners and avoid trying to provide everything for the neighborhood and the school itself. Because of the diversity of the partners, the neighborhoods, and the school communities, no two Dayton Neighborhood School Centers are alike. While this is difficult to manage, it is consistent with developing the schools as vital centers of their diverse neighborhoods.

Site coordinators

Each of Dayton's Neighborhood School Centers has a full-time site coordinator who is employed by a nonprofit agency partner, with supervision provided by the agency and the school principal. Coaching is provided by two consultants from the Fitz Center: one for support on building community in the neighborhood and one for assistance with building community within the school. Site coordinators are selected by a team of partners representing the school system, the school building, the nonprofit agency partner, and the Fitz Center.

The current site coordinators were selected primarily from outside the school community. Most are young and energetic, with limited experience. One is a former teacher. All are skilled community builders who use consensus organizing techniques to add value to the schools as centers of their neighborhoods.

The site coordinator job description is illustrative of the primary responsibilities, but daily flexibility is essential:

- Serve as broker between school and community
- Assist teachers with coordinating classroom activities funded by Neighborhood School Centers
- Implement service-learning projects with area colleges and universities
- Develop at least one new relationship with a community partner each quarter and maintain relationships with current partners
- Assist with coordinating an active community education council and steering committee
- Coordinate workshops for parents and staff
- Act as a team member by assisting with various duties, when available, such as cafeteria and recess duty or clerical stand-in

The site coordinator supports the school principal by managing the contributions of various community partners who bring programs voluntarily to the school site. The site coordinators are employed by leading Dayton nonprofit organizations committed to the vision of making schools the centers of their neighborhoods. The site coordinators and their employers are the brokers of partnerships for the Neighborhood School Centers. Three paid interns from UD assist each site coordinator and help bring other university students and faculty to the school site as service learners.

Neighborhoods as partners

While it is difficult to create and sustain partnerships between schools and neighborhoods, it is vital to our community school concept. Leaders of neighborhood associations, community development corporations, churches, parent organizations, and neighborhood hospitals have participated in the planning and implementation of Dayton's Neighborhood School Centers. With the assistance of the City of Dayton Planning Department, the neighborhoods have joined forces to apply for Ohio Department of Transportation Safe Routes to School funding for the NSC neighborhoods. For three of the school sites, neighborhood planning teams participated directly in planning new school buildings and locating them to maximize their impact on the future of the neighborhoods. The social capital, both bonding and bridging, developed during the building planning period exceeded anything that could be expected from any other organizing strategy. These neighborhoods even feel ownership of school buildings that are yet to be built.

The neighborhoods in Dayton's Neighborhood School Centers project have both great assets and serious needs. For Dayton's Neighborhood School Centers to be successful, these neighborhoods must move forward. Each neighborhood has identified one or more key developmental assets on which to build. These include a national park site in the neighborhood that was home to the Wright brothers, thirty home-building projects in two neighborhoods, a new family center funded by the Kroc family, a comprehensive neighborhood redevelopment project around one school, city parks, stable neighborhoods, hospitals, and strong neighborhood associations.

Barriers to a hopeful future will have to be removed. These include extensive poverty and little economic diversity, high foreclosure rates, low student academic achievement, youth crime, and parent preference for busing over walking to school. Assets must be leveraged to overcome barriers and ensure that the neighbor-

NEW DIRECTIONS FOR YOUTH DEVELOPMENT • DOI: 10.1002/yd

hood and its citizens continue to work for the school as the center of the community.

Private sponsor partners

Recruited and organized by the Dayton Foundation, sixteen sponsoring partners share a commitment to the vision and objectives of Neighborhood School Centers and provide funding to plan, organize, administer, and evaluate the pilot initiative. Their motivations range from concern for the common good, to children's advocacy, to enlightened self-interest. Recruiting them to the project was the personal mission of the Dayton Foundation executive director, community projects director, and board chair. All sponsors committed to three years of support for the pilot project. Other than United Way and UD, none was asked to continue funding beyond the pilot demonstration. The private funding sources are representative of a Dayton community united behind an important idea.[8]

While funding from most of these partners will end in 2009, some have initiated targeted funding for certain aspects of the Neighborhood School Centers. The Iddings Foundation is piloting a $900,000 Positive Behavior Support Initiative at four of the schools; the Dayton Foundation made a separate grant award to support arts programs; Children's Medical Center and Kohl's provided bike helmets at one site; and UD and United Way are expected to continue direct support beyond the pilot project.

Public partners

While Dayton's community initiatives are frequently stimulated by grants from private or public sources, sustainable efforts almost always have an identified public sector partner. Besides the access these partnerships provide to community-wide constituencies, public funding provided to projects from city, county, state, or federal sources is key to sustainability. Local experience is that most private sector partners invest for three years to stimulate creative problem solving and then move on to other worthy and interesting initiatives.

Without sustainable sources of funding, community programs do not build community. Youth programs, in particular, risk leading impressionable young citizens on a roller-coaster ride of programming. This can happen due to the "three year and out" approach of private sponsors or the inconsistency of fiscal federal block grant funding. Either way, children, youth, and adults alike see summer programs, after-school opportunities, and youth employment programs as occasional rather than as part of the neighborhood community fabric. By always blending private and public funds available primarily from or through local sources, Dayton's Neighborhood School Centers hope to establish predictable neighborhood assets.

As start-up financial support from multiple private sources ramps down in years four and five, Dayton's key public partners will play a larger funding role. Each has been involved from the start, and they participate in a variety of ways, not all financial. Dayton Public Schools has contributed funding to the pilot project and will assume a greater funding role in years four and five. The superintendent, deputy superintendent, members of the board of education, and principals are directly involved in the planning and implementation of Neighborhood School Centers. The superintendent hosts a monthly meeting of the partners, including city, county, lead nonprofit executives, and Fitz Center team members, with the expressed mission of removing barriers to achieving the shared vision.

Dayton Public Schools also enabled the Neighborhood School Centers planning committees at three sites to plan the new school buildings. Architects and planners joined teachers, parents, neighborhood leaders, nonprofit partners, and business owners to site the schools and design the buildings within the guidelines and cost allowances of the Ohio School Facilities Commission. In addition, having failed to pass one operating levy since the start of

Neighborhood School Centers, Dayton Public Schools used the early successes of the project to convince voters to approve a subsequent levy issue. The Neighborhood School Centers won some broad-based community support for the levy effort and provided tangible evidence that the school board and administration are committed to neighborhood schools as community centers. The superintendent of Dayton Public Schools has become a champion of Neighborhood School Centers and has committed to continue funding almost half of the project costs in years four and five.

The City of Dayton has not been a direct financial sponsor and is not likely to become one. The city's roles are primarily three: property acquisition, recreation, and safe walks to school. City planners have been active in the siting and design of each neighborhood school. Where vital to securing the right footprint for the neighborhood, the city has assisted with land acquisition and trading of properties. The community has not used eminent domain at any of the sites. At one of the school sites, the city has engaged in a significant redesign of a pool facility to create a planned splash park and other park features for recreational use. At another site, a park adjacent to the school was incorporated into the school as playfields. At a third, the school and nonprofit partner are using a nearby park as if it was the school playground. All three arrangements work for the schools and neighborhoods. In addition, the City of Dayton Department of Planning is leading community efforts to secure a significant Safe Routes to School grant through the Ohio Department of Transportation.

Montgomery County has been a funding source from the beginning. While most of the funding to date has been through Temporary Assistance for Needy Families, the Montgomery County Family and Children First Council recently voted to pilot a comprehensive neighborhood initiative in one or two neighborhoods. This state-mandated body is seeing the Neighborhood School Centers as a strategic tool to combine efforts to produce two of the county's desired outcomes: successful youth and safe and supportive neighborhoods. Some of the key nonprofit partners in the project also receive project grants from the discretionary portion of the human services levy. Montgomery County's ongoing support is key to stable and sustainable programming for children, youth, and families. Neighborhood programming is not a traditional approach to most human and social services in the county. The Neighborhood School Centers initiative hopes to demonstrate the importance of neighborhood context to positive human and social outcomes.

The Dayton Foundation role

The Dayton Foundation organized the Neighborhood School Centers project, provided some of the funding, and raised the rest from other foundations, corporations, hospitals, and UD. The foundation has relatively small discretionary funding sources, so this project was a major commitment of human and financial capital. Its executive director, community projects director, and board chair have been directly involved in the project for five years, including two years of planning and three years of implementation. The foundation is rightly recognized for its leadership role in bringing the community together around this important initiative.

University of Dayton leadership

The West Philadelphia Improvement Corporation (WEPIC) replication project taught the University of Dayton how to do the work required to link a university, neighborhood, and school. The University of Pennsylvania supported early work with Patterson-Kennedy Elementary near the UD campus beginning in the mid-1990s. The partnership with UD students and faculty went beyond the School of Education and Allied Professions and led to involvement by many other university departments and individuals in a neighborhood school. The University of Pennsylvania showed UD how to do this by demonstrating its own methods of organizing in the middle schools of West Philadelphia. Patterson-

NEW DIRECTIONS FOR YOUTH DEVELOPMENT • DOI: 10.1002/yd

Kennedy's cybercafé was the first major achievement, but many other projects evolved from the partnership over the next decade.[9]

Ironically, Patterson-Kennedy is a very old school in a neighborhood that no longer has many school-aged children, and it will be closed after the other neighborhood schools in Dayton are opened. But UD has learned a lot of lessons at Patterson-Kennedy, and success there led directly to the university's participation in the Neighborhood School Centers project. Continuing support from Penn connected us to the community school movement and the Coalition for Community Schools. The community school model was a very good idea at an important time in Dayton's history.

It is safe to assume that UD could not have accepted the leadership role on the Neighborhood School Centers project without the Fitz Center. Although the center is only six years old, its work builds on that of a predecessor organization that existed from the mid-1970s until it was incorporated into the Fitz Center in 2002. That organization, funded largely by the Society of Mary, which founded and operates UD, helped organize many of Dayton's fifty-five neighborhoods and multiple neighborhood development corporations and community gardens. This community organizing experience is now part of the Fitz Center and gives the center credibility in the neighborhoods. The Fitz Center namesake, Brother Raymond Fitz, S.M., and most of the professional staff are experienced and trusted community leaders with track records as consensus organizers and facilitators of large-scale, high-profile community projects.

The capacity of the Fitz Center to convene, organize, motivate, facilitate, and manage what amounts to a confederation of stakeholders seeking to become partners with a shared vision and mission is key to the university's role in this and other community initiatives. Many individuals, departments, centers, and programs of the University of Dayton are active in the Dayton community. However, the community recognizes only a few as the brokers of its interests on campus. The Fitz Center is one of the community brokers and, as such, maintains an active working relationship with the university president and other key administrators whose work in the community is significant but secondary to their leadership of the university. The Fitz Center agenda is largely determined by the needs and aspirations of the Dayton community. By allowing for such a center within the academy, UD makes community partnerships manageable, sustainable, and valuable for both the university and the community. UD's Urban Teacher Academy assigns first- and second-year teacher education students to the Neighborhood School Centers for field experience. The goal is to retain graduates in Dayton Public Schools at all grade levels.[10]

Indicators of success

It is premature to declare Dayton's Neighborhood School Centers either successful or failing. Because of the importance given to sustainable relationships, it will take five to ten years to draw a definitive conclusion. However, there are already indicators that look more like success than not:

- *Multiple programs have been started at each site.* They reveal the breadth of school and neighborhood interests and the variety of partners that have been attracted. Youth programs are the most common. They vary from one center to another, but all emphasize youth and community assets. More than forty projects and programs have been introduced.
- *The University of Dayton School of Engineering has initiated a robotics program at one center, with plans to expand to others.* Communication skills are being developed among seventh- and eighth-grade students at one center through a school Camera Crew and PhotoBook, both extracurricular activities. Spanish-speaking students and their families are assisted at two centers by UD students fluent in Spanish and the creation of a club for students to work on English language skills in an enjoyable after-school atmosphere. A premier Dayton after-school program, Adventure Central, brings its highly respected programs focused on nature to one center twice weekly. Gender retreats enable

students at another school to gather on the UD campus to hear from and interact with community leaders of their same sex. One center has organized a service club that takes students into the community as providers of services rather than as receivers. At all Neighborhood School Centers, The YMCA has introduced intramural sports teams for fourth-, fifth-, and sixth-grade students. Team sports had disappeared from Dayton's urban neighborhoods and public elementary schools.

- *The Neighborhood School Centers project has captured the imagination of community leaders, city planners, and local foundations.* The centers are attracting resources focused specifically on the Neighborhood School Centers and the surrounding neighborhoods. Grants are being written for external support that use the centers as the proposed means of concentrating effort to demonstrate the potential of new ideas. This leveraging of resources is one of the benefits of community schools.

- *The Neighborhood School Centers are attracting or supporting community investments in new housing.* At the future site of one Neighborhood School Center, a major neighborhood redevelopment project is under way with the school as its centerpiece. The new commons will include the school, park, playing fields, walking paths, and a pool and splash park. Thirty new homes being built in the neighborhood will be rented with an option to purchase. Neighborhood safety has already improved through the use of neighborhood policing. At another site, urban youth who are enrolled in a builder's academy have already completed thirty-three new homes for purchase and are continuing to build and sell homes. A third neighborhood has developed a housing plan to support comprehensive neighborhood development.

- *Sustainable relationships characterize the Neighborhood School Centers programs and partnerships.* The relationships are based on mutual self-interest and, as such, are more likely to be sustained than one-way service relationships. Why else would partner organizations bring their own limited resources to schools that are Neighborhood School Centers? The project must continue to recruit new partners to each site, but the potential of focusing an agency's efforts at a single school as part of that agency's core service mission has been demonstrated.

- *One successful charter school has been folded back into Dayton Public Schools.* As a school with site-based management, this Neighborhood School Center is set to demonstrate the full potential of the community school model. Neighborhood leaders and a strong nonprofit partner are more directly involved in the leadership of the school, including selection of the principal. This model, if successful, could lead to similar experimentation at the other Neighborhood School Centers.

- *Parent choice of neighborhood schools over other options is slowly increasing.* At the new Ruskin School, more than one-third of the students were walking to school in fall 2008. This is the hoped-for direction for all of the Neighborhood School Centers. No one in Dayton is required to go to a certain school, so neighborhood schools must become schools of neighborhood choice. The Fitz Center and Dayton Public Schools are tracking each of the Neighborhood School Centers to monitor enrollment from within one and a half to two miles of the school. The one-and-a-half-mile radius serves as an indicator of potential walkers.

- *The project has successfully brokered additional resources for the school community and students.* Having a person with the designated role of brokering relationships between the school and the community (especially the UD community) has garnered a large array of volunteer hours, in-kind donations, monetary donations, and grants that have supplemented and supported new and creative opportunities for academic support, learning, and self-discovery for the students in the school. Future funding for years four and five has been secured sufficient to sustain current levels of activity at five Neighborhood School Centers. The total annual cost of Dayton's Neighborhood School Centers is $400,000. Future funding sources are Dayton Public Schools, Montgomery County, United Way of the Greater Dayton Area, and UD. Dayton Public Schools will pay approximately half of the total cost in years four and five.

- *The site coordinator is the key role and most important investment.* The Neighborhood School Centers project has resulted in the successful creation of a model that has been implemented in five schools and neighborhoods. The project model is of a site coordinator with a foot in the neighborhood community and a foot in the school, with the goal of bringing the best resources to bear on the children and the school for the promotion of student academic success, creating a neighborhood hub of activities, and creating such an inviting school that parents choose this model for their children. This model has proven useful as one that contributes to school and neighborhood and is viewed as a desirable opportunity by schools not in the project.

Challenges we still face

Many challenges remain for Dayton's Neighborhood School Centers. Can we help contribute to improved student academic performance? (Most schools are in Academic Emergency status, according to Ohio performance standards.) Can we take the Neighborhood School Centers to scale? (Dayton has twenty-two public elementary schools.) How will a larger project be administered, and by whom? (The guiding methods of the Neighborhood School Centers could be emphasized in various ways, depending on the administrative structure.) Will Dayton parents choose to have their children walk to a neighborhood school after a generation of busing them to the school of their choice? (In all but one of the Dayton neighborhoods served by the Neighborhood School Centers, virtually no child walks to elementary school.)

In spite of these challenges, no one has walked away from a leadership role on this project. The sponsors and partners remain believers. UD remains committed to a leadership role. And the Fitz Center continues to believe that Dayton's Neighborhood School Centers are the best demonstration of the power of community building in urban Dayton. Patient persistence may be the most important

NEW DIRECTIONS FOR YOUTH DEVELOPMENT • DOI: 10.1002/yd

leadership trait needed to realize a hopeful future for the youngest citizens.

The possibility that smaller is better

Dayton is as hard hit as any other urban center in the United States. If the Dayton Neighborhood School Centers succeed in realizing their mission and shared vision, the importance of programming within neighborhood scale and leveraging strong personal relationships will have been demonstrated in a new way. If so, the lesson of Dayton may be that smaller is better.

One of the biggest challenges youth programs face in our urban community is inconsistency. In Dayton, our youth have experienced a roller-coaster of programming throughout their young lives. Depending on the availability of federal and foundation grants and the resulting large fluctuations in local youth program opportunities, these young people have experienced school years and summers with interesting and even exciting programs followed by years of nothing. Youth sports teams and leagues, as just one example, have not survived the inconsistent support and funding.

Our community is developing its Neighborhood School Centers to provide predictable and sustainable programs for youth, families, and neighborhoods. We are using partnerships of public elementary schools with effective nonprofits to build program capacity that can be counted on year to year at the neighborhood level, with or without external grant support. To accomplish this, we are starting small, moving ahead methodically, and emphasizing the strength of personal and institutional relationships.

Notes

1. Kretzman, J. P., & McKnight, J. L. (1993). *Building communities from the inside out.* Chicago: ACTA Publications.

2. Putnam, R. D. (2000). *Bowling alone: The collapse and revival of American community.* New York: Simon & Schuster.

3. Senge, P. M. (1990). *The fifth discipline: The art and practice of the learning organization.* New York: Currency Doubleday.

NEW DIRECTIONS FOR YOUTH DEVELOPMENT • DOI: 10.1002/yd

4. Heifetz, R. A. (1994). *Leadership without easy answers*. Cambridge, MA: Belknap Press.

5. Eichler, M. (2007). *Consensus organizing: Building communities of mutual self-interest*. Thousand Oaks, CA: Sage.

6. If there is a skill that we would hope to continue to improve in the Fitz Center, it is the ability to facilitate dialogue. Helping groups learn to come together and suspend judgment long enough to fully digest each other's viewpoints has proven critical to the community building process. Conversation circles have not won general acceptance yet, but the center continues to begin most of its projects with open, nonjudgmental dialogue. It is usually where we start identifying the potential for a shared vision. Without this vision, community schools and most other community initiatives are usually short-lived and ineffective.

7. Putnam (2000) aptly distinguishes between bonding and bridging social capital: "Bonding social capital is good for undergirding specific reciprocity and mobilizing solidarity. . . . Bridging networks, by contrast, are better for linkage to external assets and for information diffusion. . . . Bonding social capital constitutes a kind of sociological superglue, whereas bridging social capital provides a sociological WD-40" (pp. 22–23).

8. This work included three programs funded by the Corporation for National Service. Fitz Center staff managed and evaluated Ohio's first AmeriCorps program for seven successful years (1994–2001). It was a collaboration of thirty agencies organizing seven community-based projects over a year. The University of Dayton and Patterson-Kennedy (PK) share a neighborhood in transition and a commitment to community involvement. CHESP results included a lead paint awareness campaign, tutoring, English as a Second Language support, after-school programs, physical education for special-needs classes, parent and teacher workshops, concerts, playground upgrades, marketing and public relations, citizenship lessons, and mini-grants to Patterson-Kennedy teachers for K–6 service-learning. By 2004, the positive effects of the UD-PK partnership (a model for Neighborhood School Centers) were evident at all grade levels. At PK, 150 students had participated in service or community-building lessons; overall activity drew in 35 PK teachers; 50 UD faculty, staff, and graduate assistants from 20 twenty departments; and an average of 175 UD students a year. Although funding ended, the partnership remains intact, with many UD students still involved at the school. Most recently, UD has been part of Ohio Campus Compact's AmeriCorps*VISTA program. Two VISTA volunteers were invaluable in launching UD's Semester of Service program (twenty students a year engaged in full-time service coupled with a three-credit course). The Fitz Center's current VISTA volunteer is working to extend the curriculum of Adventure Central, a respected after-school program, to the Fairview Neighborhood School Center, tying in university student support. Fairview and the other Neighborhood School Centers will directly benefit from the experience.

9. The sixteen private organizations and foundations that funded planning and start-up of Dayton's Neighborhood School Centers represent a diverse and generous philanthropic community: the Antioch Company, the

Children's Medical Center, Culture Works, the Dayton Foundation, DP&L Foundation, Fifth Third Bank, Grandview Medical Center, the Iams Company, Iddings Foundation, KnowledgeWorks Foundation, MeadWestvaco Foundation, NCR Corporation, Premier Health Partners, United Way, University of Dayton, and the Virginia Kettering Foundation.

10. Two other UD assets are important to recognize. The Business Research Group within the Center for Leadership and Executive Development of the School of Business Administration is a vital research support to Dayton Public Schools and the Neighborhood School Centers initiative. It provides current and accurate school demographics, survey research, and needs assessment from which everyone works. The School of Education and Allied Professions, its dean in particular, is the most important university partner with the Dayton Public Schools and with all aspects of the education of Dayton's urban children and youth. Besides wise counsel and solid teacher and administrator preparation, the school has developed an Urban Teacher Academy and the Dayton Early College Academy (a high school); has supported urban charter and Catholic schools in their development; has moved science, technology, engineering, and math education to the forefront in the Dayton economy; and has pioneered many of the area's best practices in early childhood education. The Neighborhood School Centers project also has benefited from the work of a loaned consultant from the School of Education and Allied Professions. The consultant is experienced and knowledgeable and a long-time confidant of Dayton Public Schools administrators. Besides helping those from outside the system navigate their way, the consultant serves as coach to the site coordinators and assists them in becoming valuable players in the school setting.

DICK FERGUSON *is executive director of the Fitz Center for Leadership in Community at the University of Dayton.*

University presidents play a vital role in encouraging university involvement in democratic partnerships. Despite the significant challenges, such efforts enable universities to best fulfill their academic and civic purposes.

6

The president's role in advancing civic engagement: The Widener-Chester Partnership

James T. Harris III

WHEN I THINK BACK to 2002 and my first month as president of Widener University, two meetings remain etched in my mind—not so much because of what happened in those meetings but rather the symbolism behind what the very scheduling of such meetings for a new president would come to represent.

The first meeting was with a group of Widener administrators to discuss plans to build an eight-foot-high fence with one entrance gate around the freshman dormitories on our main campus in Chester, Pennsylvania. I do believe that fencing can serve as both a means of securing a campus while adding aesthetic value, but I thought that building a fence with a gate would give the perception to our neighbors that they were not welcomed on campus, and it might exacerbate the already strained town-gown relations. When I rejected the plans and suggested that by working with the

community in deeper and more meaningful ways we could secure our campus better, the response from the group was unenthusiastic. As one vice president stated in response to my suggestion for greater community engagement, "That will never work. Chester is a city that will suck you dry."

The second meeting was one that was planned by the university relations office and a local newspaper in Chester. The scheduling of the meeting was appropriate, but what was interesting was that it had been arranged for me to be picked up by a campus security guard in a marked car that morning to be delivered to the newspaper office less than a mile away. When I informed the person who had scheduled the meeting that I would drive myself, I was told that it was too dangerous for me to drive into Chester alone and that she could not be held responsible if anything happened to me.

These two stories reveal how some administrators at Widener viewed the relationship between the university and city of Chester in summer 2002. Although some good partnerships had developed over the years, the negative attitudes that I encountered were born of years of frustration from a university with limited resources that was attempting to engage a city that was increasingly difficult to deal with due to all of the challenges associated with an urban community in decline.

From the beginning, it was clear to me that any efforts to engage organizations in Chester in meaningful and democratic partnerships would require much time, effort, and considerable university resources. Along the way there have been questions raised about how much such partnerships "cost" the university and what the return on our investment is. My response has been that there are costs associated with whatever type of relationship you decide to have with your community.

As we moved forward with a bold plan to become a model for university-community democratic partnerships, it would become clear that while there are additional costs associated with pursuing this course of action, it would be no more of a cost than building fences or isolating ourselves from the harsh realities just outside our campus gates. In fact, what we would learn is that by creating democratic partnerships and not caring who received the credit, we could strengthen our academic offerings, make our campus more secure, and develop trust among key community stakeholders.

The Context: The City and the University

The City of Chester is located ten miles southwest of Philadelphia in Delaware County, Pennsylvania. It has a proud history dating back to 1644 when it was founded by Swedish settlers. In 1682, William Penn renamed the small settlement "Chester." Chester played a prominent role in the early colonies, and by the twentieth century, it had emerged as one of the nation's leading industrial cities.

By the 1950s, the city's population had swelled to sixty-six thousand, mostly due to the significant ship building and manufacturing that prospered during the two world wars. It was during this time that Martin Luther King Jr. attended Crozer Seminary, earning his degree in divinity and serving as an associate pastor in a local church. Chester had many points of pride and had clearly earned its motto: "What Chester makes, makes Chester."

By the 1960s, the city began experiencing economic difficulties as manufacturing and other industries moved out of the city. By the time of the new millennium, it faced the challenges of an urban environment in decline. By 2000 the city population had dropped to under thirty-seven thousand, with more than 27 percent of all individuals categorized as living in poverty and 41 percent of the adults listed as outside the labor force.[1] By 2007 the median family income in Chester was $31,928 as compared to $51,170 nationally and $62,223 for the surrounding county.

As problematic as economic growth and community development were in the City of Chester, the challenges facing the public schools were equally daunting. The Chester-Upland School District (CUSD) is one of the most troubled school districts in the nation. Out of 501 school districts in the Commonwealth of Pennsylvania, the CUSD has ranked at or near the bottom for over two decades. The ability of Widener or any other university to work with the school district in meaningful ways over the years was always compromised by the lack of consistency in district leadership. Over the past decade, at least eight different people had served in the position of superintendent.

Widener University has a long and proud history as well. Founded in 1821, the school, which would become known as the Pennsylvania Military Academy, moved to Chester in the middle of the nineteenth century. It changed its name to Pennsylvania Military College in 1892, and in 1972 when the corps of cadets was retired, the school became known as Widener College. In 1979 it earned university status, and by 2002 had developed into a multi-campus doctoral-granting institution with four locations serving approximately sixty-five hundred students through eight colleges and schools.

The main campus in Chester attracts most of its student population from the greater Philadelphia metropolitan region. The undergraduate full-time students are predominantly Caucasian and drawn mainly from middle-class families, with 28 percent of those students being Pell Grant eligible. The university's graduate professional programs attract mostly working adults from the region.

Over the years, Widener's predecessor institution, Pennsylvania Military College, was viewed as a critical part of Chester. Many alumni tell stories of marching through town to participate in the opening of a new store or to be part of some local celebration. In the 1960s, the college started a precollege program entitled Project Prepare to help inner-city youth succeed in college, and the success of that program fostered goodwill between the city residents and the institution.

Nevertheless, the 1960s were tumultuous times in Chester. As the city declined, crime rates grew, further damaging the city's reputation. When the tax base narrowed and the middle class migrated to the suburbs, Widener was caught in a difficult situation. As a growing university with limited resources, it needed to be careful in its investments, so as things started to deteriorate in the city, the university started developing a citadel mentality and warned its faculty and students of the dangers that lurked just beyond the walls of the academy.

What drove a further wedge between the university and the community was that as the public schools deteriorated and property taxes rose, fewer Widener faculty and staff were choosing to live in the city. Whereas during the more prosperous days of Chester the majority of Pennsylvania Military College employees lived in the city and participated in local issues, by the year 2000, less than 10 percent of Widener employees lived in the city limits. All of these factors, as well as decisions made by the university, led local citizens to start viewing Widener as an institution that was concerned only about promoting its own self-interest. As the mayor said to me; "Widener is viewed by most citizens as a dragon that eats up land that otherwise would be generating tax dollars for the city."

Although there was no university-wide plan in place by the turn of the century to strategically engage the local community, Widener had created a partnership with the Crozer-Chester Medical Center and a few other organizations to establish University Technology Park, a nonprofit corporation designed to attract high-technology firms to Chester. Unfortunately, within a few years, the park was only half occupied, and only Widener and Crozer remained as partners.

Around the same time, the Center for Social Work Education partnered with the Chester Education Foundation to establish Social Work Consultation Services (SWCS) in 2000. From it inception, the SWCS articulated a dual mission: to improve the lives of low-income citizens in Chester and train competent and caring social work leaders.[2] Although this new entity was well received in the community, the university administration did not publicly support its creation, leaving the faculty to raise the resources to manage SWCS on their own.

By 2002, Widener was viewed as a university located in a troubled urban environment, with no strategy in place for systematically addressing the significant issues facing Chester. It was clear we needed to rethink our relationship with Chester and develop a meaningful vision forward that would direct our work as a university. For years, Widener had operated on an annual operating plan that drove the budgeting process. According to the records, it had not engaged in a strategic planning process in over a decade, and there were no long-term university plans in place in 2002. The mission statement was common and uninspiring and essentially gave the university wide latitude to meet the demands of the marketplace.

Strategic planning for civic engagement

In the fall of 2002, we established a university strategic planning committee capably chaired by Joe Baker, vice president for finance. In my first months on the job, I had met with over a hundred trustees, faculty, and staff, as well as dozens of community and alumni leaders. In each meeting with internal constituents, I asked who they thought should serve on a strategic planning committee to direct the future of the university, and from that information I asked twelve people to serve. We also were fortunate to have the services of Rod Napier as a planning consultant to help us shape our decision-making processes.

The Strategic Planning Committee (SPC) set out to create a dynamic new vision for Widener starting with a new mission statement. The mission statement was written by a small group of faculty leaders, trustees (skillfully led by chairman David Oskin), and key senior staff members. The mission, adopted by the board of trustees in December 2003, boldly stated that Widener would "connect the curricula to societal issues through civic engagement" and that we would "contribute to the vitality and well being of the communities we served."[3] The strategic plan had several goals, including one that specifically stated that the university should address the metropolitan region's most pressing problems. The mission, strategic goals, and vision statement, entitled "Vision 2015," were approved by the board at its May 2004 meeting.

When the planning process was completed, over twelve hundred people had participated, representing all university constituent groups and including several local citizens. Dozens of meetings took place with elected officials, business leaders, clergy, community activists, members of the Chester-Upland School board, and neighbors.

As part of the document, the vision statement proclaimed that Widener would become the "nation's preeminent metropolitan university" within a decade.[4] It is important to note that the designation of Widener as a "metropolitan university" was a new direction for the university. Most people had not heard of the term, and some saw it as possibly limiting the scope of the university's potential. However, most saw the potential for the university to make its mark nationally by focusing locally on important issues. When the final plan was developed, our focus on the metropolitan region could be summarized as concerning three critical areas: community development, economic development, and public schools.

While everyone who participated in the planning process shared a common belief that Widener had the potential to achieve new levels of academic distinction, the prevailing attitude about the university among most Chester community leaders was skepticism. They were skeptical about whether Widener had the ability to lead or even participate in a meaningful way in a renaissance in the city, as well as if they had the fortitude to take on some of the toughest issues, especially those dealing with the public schools. Here, I focus on one of the three areas identified as critical in our plan: addressing the needs of a failing school district and creating meaningful and sustainable partnerships.

When we began publicly discussing our new mission, the response was overwhelming. From the first, many students and faculty decided to encourage community service as much as possible as an expression of our mission. Whereas prior to my arrival, students were warned in freshman orientation "not to cross over the bridge" (this refers to the bridge over Interstate 95 connecting the campus to the downtown), "crossing over the bridge" into the heart of Chester became the new mantra for many. However, their enthusiasm to volunteer in Chester was dampened when many volunteer groups found that few social service agencies had the organizational capacity to handle large numbers of student volunteers.

A second problem emerged because few of the student groups that did find volunteer opportunities in Chester were trained appropriately to deal with what they would encounter, and many completed their service with no opportunity to reflect in any formal way about their experience. This led to some exploring by faculty about ways to create more meaningful learning experiences tied to academic programs.

Fortunately, a small group of faculty, led by senior faculty member Arlene Dowshen, had been meeting on a regular basis to discuss the possibility of incorporating service-learning into the curriculum. After my first year, I realized that for the university to truly engage the community and take advantage of the willingness of a few faculty members to create community-based learning opportunities for students, we needed to start coordinating these efforts.

We were able to attract Marcine Pikron-Davis to Widener in 2003 as my special assistant to coordinate these efforts. Within a year, we secured funding for a service-learning fellowship program for faculty. Dowshen and Pikron-Davis worked with the provost, Jo Allen, to design a program so that faculty could receive a one-course semester release for one year to develop and offer a community-based learning experience. From 2004 to 2008, forty-nine faculty participated in these seminars, and seventy courses were developed that focused on issues in the local community, with many aimed at the school district. To date, between twelve and twenty service-learning courses are offered every semester, and over twelve hundred students have taken these classes.

In addition, through the leadership of Stephen Wilhite, dean of the School of Human Services Professions, Widener became a member of the Coalition of Urban and Metropolitan Universities (CUMU). This involvement helped Widener faculty and staff to form relationships with other metropolitan universities and to discover best practices in community engagement (with special emphasis on public schooling initiatives) at similar institutions. This desire to discover best practices led to several visits by members of the senior administration to other colleges and universities that had a reputation for innovative approaches to community engagement. Fortunately, we did not need to go far, for one of the best examples in the country is the University of Pennsylvania and its work in West Philadelphia. Ira Harkavy, director of Penn's Center for Community Partnerships, and several of his colleagues embarked on key community building initiatives that were both bold and innovative. Positioning these efforts within the context of the democratic responsibilities of higher education, Penn emerged over the previous decade as the leader in this movement.[5] However, there were other good examples, and over the next two years, members of the university's senior leadership team made visits to several institutions across the region. Through these visits, we developed a comprehensive and integrated civic engagement model while simultaneously creating strategies for addressing issues concerning the Chester Upland School District.

One of the most significant challenges to creating democratic partnerships is the incapacity of most nonprofits to absorb a large number of volunteers or properly use the resources universities might provide to tackle community concerns. To address this issue, we supported the efforts of the Social Work Consultation Services to help strengthen the internal capacity of nonprofits to better use student volunteers and find ways to benefit from the growing number of community-based learning opportunities developing at Widener. We also discovered that there were few neighborhood organizations in place to help citizens work with the university or the city to air their concerns about issues. To that end, we created a leadership program where local citizens could receive training at the university on how to organize to address significant issues. To stimulate an ongoing dialogue between the university and its neighbors, the Sun Hill Coalition was also created and included local citizens, students living off campus, and university officials.

During this time Widener was invited to participate as a member of Project Pericles, a nonprofit organization whose stated mission is to encourage and facilitate commitments by colleges and universities to include education for social responsibility and participatory citizenship as an essential part of their educational programs, in the classroom, on the campus, and in the community. Membership in Project Pericles required the university board of trustees to create a new board committee that focused directly on civic engagement. The board unanimously endorsed the idea, and several key board members were named to this new committee. What this did was establish civic engagement as core and central to the work of the board.

As positive as these efforts to engage the community were, one major area of concern to most people was consistently brought to our attention when we discussed with local citizens and elected officials how we should be engaging the community: the role Widener could play in helping the commonwealth's worst school district. The big question for the university was not whether Widener could engage the Chester Upland school district in meaningful ways but whether our efforts could be sustainable. Our journey to discover something that was equally meaningful and sustainable would take us in many different directions and lead us to a conclusion none of us would have imagined when we began.

The promise and pitfalls of university-school partnering

The Chester Upland School District serves slightly under five thousand students from kindergarten through twelfth grade. In 1994, CUSD was identified by the Commonwealth of Pennsylvania as the worst-performing school district in the state. Its multimillion dollar deficit, coupled with the poorest performance by students on standardized tests in the commonwealth, led to a state takeover of the district. The district remained under state control until 2007, when that designation was lifted in spite of continued poor fiscal management and lack of achievement of students. For example, in 2006–2007 only 9 percent of all eleventh-grade students passed the statewide proficiency test in language arts and only 3 percent passed the math proficiency test.[6]

By the time I arrived in 2002, a for-profit company, Edison Schools, had been retained by the CUSD board to try to improve the struggling district. Edison was responsible for running eight out of the nine public schools in the district while the CUSD remained responsible for managing the buildings and transportation of the district's children. Furthermore, since 1997, the school district had witnessed an unprecedented growth in charter schools that came into effect after the state passed legislation for the establishment of "self-managed public schools," as defined by the state Department of Education. In Chester, most of the children in charters were enrolled in schools run by a for-profit organization that had ties to the political leadership in the county.

Early in my tenure, I was introduced to Edison's CEO and superintendent, Dexter Davis. We decided we would meet every two months to see if we could find ways for Widener and CUSD to work more closely together. Davis was very interested in establishing close ties with Widener, but given that he did not have direct control over the operations, curriculum, or teachers in eight of nine buildings in the district or the charter schools, his ability to facilitate a closer relationship was limited. In my mind, I was hopeful that Widener might be able to work with the CUSD to create a school district–university partnership similar to what the University of Pennsylvania had established with the Philadelphia public schools. Although we did not possess the same ability for funding or leverage as Penn, I thought its model, as well as several other national models that were emerging, might provide ideas for ways in which Widener could partner with the CUSD in a meaningful manner.

Over the course of that first year, Davis and I attempted to collaborate on several projects. The most notable was an effort to raise funds to provide a new computer lab for the middle school located directly across the street from the university. By December Widener had raised fifty thousand dollars to purchase new computers, desks, and learning software, and all the school district needed to do was provide the room setup and the service to support the new lab. Of course, all of this had to be coordinated between the district, Edison, and Widener. Due to the complicated and awkward relationship between the district and Edison, the lab was not ready to open until late May. As excited as the students and parents were about this wonderful new lab and the developing relationship with Widener, it was clear to me that even the simplest of partnerships would be difficult to orchestrate.

The complexity of the CUSD situation became even clearer to me when, in the second year, we attempted to create a tutoring and mentoring program within the district to help children who were taking the state proficiency exams. Since the CUSD had rated at or near the bottom in every test given by the state, it appeared that the quickest way to combat this problem was to provide tutoring for the children in jeopardy of failing. Davis understood this need and supported the launching of such an effort by Widener. With his support, we encouraged Widener students to volunteer for the effort and established a university-wide policy that allowed staff to take an hour off once a week at the university's expense to tutor a child in one of the two public schools within walking distance of campus. Several Widener employees decided to sign up; however, Edison neither supported nor had the ability to handle dozens of new volunteers, and the program never got started. When we inquired as to why this initiative was not acceptable to Edison, we were told that its greatest need was for hall monitors, and if we wished to provide volunteers for that effort (saving Edison the cost), they would be happy to receive them.

By spring 2005, the CUSD control board determined that Edison was not successful in helping the district meet its goals. The continuing decline in student test scores, coupled with a number of incidents within the schools, led to the state's decision to break its contract with Edison. In addition, the CUSD released Davis from his role as superintendent.

This placed the CUSD in a difficult position: the curriculum and the books and materials were proprietary materials of Edison, so it had nothing in place to start the next school year. In May, an elected official contacted me, stating stated that the CUSD was eligible to receive as much as half a million dollars in state funding to help address the myriad issues facing the district, the most pressing being the need for a curriculum in three months, when school began.

In consultation with an acting superintendent, it was determined that Widener Education faculty and CUSD faculty would work collaboratively to develop a new elementary curriculum that would be in place for the entire school district by the fall. In the end, Widener faculty, working with CUSD teachers, created a new elementary curriculum and left over $300,000 for the new superintendent to implement it. It is important to note that most of the money spent that summer went to pay for the participation of CUSD teachers in the development of the program. After two months of intense meetings, the new curriculum was ready, and both parties were excited to work together to implement it. But the new superintendent who arrived in late August decided not to use the new curriculum and instead put the remaining funds into a curriculum she had used at another school.

During my first four years, Widener responded to a myriad of requests from the school district. In addition to the initiatives previously described, other programs were created, including a mathematics enrichment program, professional development for CUSD teachers, after-school tutoring programs, the establishment of Saturday academies in science and social sciences on our campus for CUSD students, and a science camp for middle school girls. All of these programs were created by Widener working in conjunction with the CUSD. Furthermore, the university took the responsibility in all cases for developing new revenue streams to pay for these innovative programs.

In most cases, though, once a program was established, it would be suspended or cut when a superintendent or building principal left, and my Widener colleagues and I would have to start the discussions over again. Widener responded to the call for support by the district on dozens of different initiatives, only to be left feeling frustrated each time.

It was during this time that Widener, led by the dean of the School of Human Service Professions, Stephen Wilhite, began looking at new options for creating meaningful and sustainable relationships with the CUSD. After years of attempting to work within the constraints set by the school district, we began seriously considering whether we should open a university charter school.

The decision to pursue a charter school that would be affiliated with Widener was not an easy one to make. In fact, my colleagues and I were opposed philosophically to the idea of charter schools. We were heavily invested in the ideals of the community-school initiatives that had been successful around the country and truly believed we could create a similar model in Chester if we only had stable leadership in the school district. But after years of frustration in dealing with a revolving door of district administrators and board members, we decided to apply for a charter to start a school that would model the very best practices in community schooling and give us some control over the destiny of the school.

First, we decided our school would be a nonprofit entity and would seek private funding to support our efforts and supplement the state funding we would receive. Second, we would play to our own academic strength in elementary education and open a K–5 school. We felt it was important to inform the district that we would place a cap on our enrollment at three hundred after five years so that they could plan for fifty students a year potentially not entering the rest of the district at the kindergarten level. We did this because one of the problems with the existing charters was that they would accept as many students as they could enroll, and the school district could not adequately plan for the needs of the remaining students.

To avoid accepting only the best students from the district, we decided to have a charter school that would accept children from throughout Chester by lottery, including those for whom English was a second language, as well as children with special needs. Based on what we had learned about the qualifications of some teachers in the district charter schools, we decided that all of our teachers would be certified and that their salaries would be competitive with the rest of the region. Making the salaries competitive within the region was important because we needed some incentive for talented teachers to work in one of the worst school districts in the state.

In the end we decided we would tie the goals of the school closely to the high ideals of the university. The mission of the Widener Partnership Charter School would be to develop urban elementary students with the behaviors, task commitment, and creativity to succeed in a tough urban environment. Our mission for the partnership school also included language similar to the university's mission, including a statement that we were preparing students to be citizens of character who can contribute to the vitality and well-being of the region.

To that end, our curriculum would include Spanish, art, and music, areas that had been cut from Chester schools decades earlier. We would also contribute to the learning environment and create a true community partnership school by drawing on the wealth of expertise and experience of Widener faculty and students across disciplines. Finally, we would create a board that would include representatives from key constituent groups: educators, parents, and community and business leaders.

From the beginning, the school was controversial. Those who opposed charter schools in Chester felt the university had relinquished it role as an equal partner in education and viewed Widener as a competitor for scarce resources. At one point, a group of community activists came to visit me and declared that as president, I had lost the moral high ground and therefore could no longer be looked to as a neutral arbiter when educational issues were being discussed in the city.

Nevertheless, many local citizens and community leaders viewed the university's proposal as the first real hope for reform in the district in decades. Many of our supporters felt that if Widener could be successful with its charter school, the other schools in the district would have to model our best practices. The university community rallied around this idea.

The first task was to hire a principal who shared our enthusiasm for educational reform and understood the basic tenets of community-centered schooling. We were fortunate to attract Annette Anderson to Widener; she had recently received her Ph.D. from Penn and had experience in this area. She quickly worked to hire a staff, and by the fall of 2006, the first hundred children entered the Widener Partnership Charter School.

During its first year, the school was located in an old hospital in the city, but by the second year, the university had purchased the original University Technologies Property, one block from campus, to house the school. As of fall 2008, the school had enrolled two hundred children, and the early assessment of the students' achievements have been well above the norms for Chester. Part of the reason the results to date have been strong is due to the collaborative nature of the school and the emphasis placed on engaging parents, the local community, and Widener faculty and students in the learning environment. For example, Widener's Center for Social Work provides outreach to families of schoolchildren who have been identified by the teachers as having difficulties. This holistic approach to learning and deep parental involvement have been two of the keys to the school's success.

The creation of a charter school was a bold move for a university frustrated with its dealings with a struggling school district. It created new alliances, but it also brought new and unexpected challenges. One big challenge was a change in the school board makeup due to a change in the state capital. Just as we had difficulties dealing with the local school administration due to turnover of key staff, the same was true at the state level. Within my first four years, three different people held the role of secretary of education for the State of Pennsylvania. It seemed that just when I would personally negotiate an agreement of understanding with one secretary about how Widener should proceed, a new one would be named. In 2006, the CUSD board filed suit in court attempting to put a cap on the enrollment of the charter schools in Chester, claiming charter schools were a drain on the financial condition of the district. Even after I had been assured by state officials that

Widener's charter was not being targeted, we quickly learned that we were included in the lawsuit and needed to file a countersuit.

After two years of litigation, the school district's case was found unconstitutional, and the state legislature passed a law prohibiting the capping of charter schools throughout the state. This lawsuit was another example of being put in a difficult spot in spite of the state's reassurances that it supported charter schools. However, in spite of these difficulties, the university community, as well as a growing number of community leaders, started viewing the academic success of the Widener Partnership School as a model for other schools in the district.

Over the years, I discovered that Widener was not alone in its feelings of frustration with the CUSD. Several other local colleges and universities had attempted to be involved in meaningful ways with the district over the years, only to meet with similar difficulties. In spring 2005, I invited the presidents of Delaware County Community College, Cheney University, Penn State–Brandywine, Neumann College, and Swarthmore College to join me at Widener for a discussion on how we might work together to help address the situation in Chester.

Three years later, this group created a new 501(c)(3) organization, the Chester Higher Education Council (CHEC), and its first accomplishment will be the establishment of the College Access Center for the City of Chester and all of Delaware County located in Chester. This group is unique in American higher education in that its membership is made up of institutions with very different missions, both public and private, all dedicated to providing better educational opportunities for the children of Chester. Another aspect of this collaboration that is unique is the personal commitment of the presidents who attend most of the meetings.

In the past year, several positive initiatives have been emerging in the district. A new superintendent, Gregory Thornton, has arrived, and he has brought innovative ideas that have been sponsored financially by the state. Some of these new ideas include the development of public–private partnership schools similar to what occurred in Philadelphia, and a few of the CHEC schools have found ways to develop stronger ties to the district. Although it is too early to state that a new day has dawned on the CUSD, there is cause for some optimism about the future. The bottom line, however, is that the enormity of the task at hand still remains difficult and will require years of continued progress by all parties.

One of the most challenging aspects of creating sustainable democratic partnerships within our community has been to definitively answer the question of whether it is worth all of the time, money, and effort we have committed. In other words, people want to know how much all of this has cost Widener University and whether it has had a positive impact,

To answer that question requires serious reflection regarding the mission and purpose of a metropolitan university and its relationship to the community where it resides. Unfortunately, it is possible to ignore what is happening in the community and to assume a citadel mentality, as evidenced by the alarming number of universities that choose not to engage in such work. Making a commitment to resolve some of society's toughest issues and developing democratic partnerships is hard work that requires a long-term approach and leadership who will not back down from a challenge. However, the cost of deciding not to engage in this work could be much greater than the investment associated with moving in this direction. For example, the costs of building a fence around our dormitory would have been several hundred thousands of dollars, if not in the millions. If we ignored our community and did not develop meaningful partnerships with organizations such as the local police department, we would be cut off from communications, which would make our campus less secure and would cost us more money to expand our own security force. Furthermore, how do you measure the costs of depriving college students from experiencing new and exciting community-based learning environments that often are rated as the best learning experiences by the students themselves?

When it comes to intervention with the school district and youth development, what costs are associated with being located in a community with the state's worst school district? When an entire generation of younger families, including faculty and staff, chooses not to live near the campus, that clearly has a negative influence on how involved they will be at the university and can stifle the creation of a robust learning community. This makes it harder for the university to recruit and retain talented faculty and staff and is likely to lead to more employee turnover, resulting in higher recruiting costs. By creating an alternative school option in the neighborhood near the campus, as well as increasing our capacity for preschool education, we are developing options for young families and making Chester a desirable place to live and work. This effort, coupled with a myriad of job training initiatives, helps strengthen the community and eventually will benefit the university by increasing the pool of qualified students, as well as provide the university with a trained local workforce.

In order to advance this important work, more effort must be made to create effective assessment tools to measure the impact of creating democratic partnerships. Perhaps more important, we should ask the equally provocative question, "Can we challenge the status quo and attempt to measure the costs associated with not doing this work?"

Conclusion

In this article, I have explored how one university charted a new direction to assist its community and marshaled its resources to come to the aid of a failing school district. These efforts demonstrate that any university that wishes to engage in this type of reform must take a long-term view and develop educational delivery systems that are sustainable in light of an ever changing K–12 environment. Sometimes those solutions may seem incongruent with the traditional view of a relationship between a school district and a local university. It has become apparent to me that a model that might fit in one district may not work in another, and universities that seek to have a meaningful impact on public schools need to adopt aspects of other successful partnerships and meld them with elements that make sense for their particular situation.

If higher education is to truly have a positive and lasting impact on K–12 education in the twenty-first century, we should encourage a myriad of models and reward universities for taking risks. Otherwise another generation of children in underperforming school districts will be lost. That is something America cannot tolerate and why higher education can no longer remain neutral. As former Secretary of State Henry Kissinger once observed: "History will not excuse the inadequacy of the response due to the enormity of the challenge."[7]

Notes

1. U.S. Census Bureau. (2001). *United States census 2000.* Washington, DC: Author.

2. Silver, P. T., Poulin, J. E., & Wilhite, S. C. (2006). From rogue program to poster child: A department's shaping of a university's mission. In K. Kecskes (Ed.), *Engaging departments: Moving faculty culture from private to public, individual to collective focus for the common good.* Bolton, Ma: Anker.

3. Widener Strategic Planning Committee. (2004). Widener strategic plan. In *Widener Executive Board meeting.* Chester, PA: Widener University.

4. Widener Strategic Planning Committee. (2004).

5. Maurrasse, D. J. (2001). *Beyond the campus: How colleges and universities form partnerships with their communities.* New York: Routledge.

6. State failed students. (2008, April 28). *Philadelphia Inquirer.*

7. See http://www.hrc.utexas.edu/multimedia/video/2008/wallace/kissinger_henry_t.html.

JAMES T. HARRIS III *is president and professor of education at Widener University.*

NEW DIRECTIONS FOR YOUTH DEVELOPMENT • DOI: 10.1002/yd

Index

Notes for Contributors

After reading this issue, you might be interested to become a contributor. *New Directions for Youth Development: Theory, Practice, and Research* is a peer-reviewed quarterly publication focusing on contemporary issues inspiring and challenging the field of youth development. A defining focus of the journal is the relationship among theory, research, and practice. In particular, *NDYD* is dedicated to recognizing resilience as well as risk, and healthy development of our youth as well as the difficulties of adolescence. The journal is also interested in applications of youth development to education and schools, and is a leading voice in afterschool and out-of-school time scholarship. The journal is intended as a forum for provocative discussion that reaches across the worlds of academia, service, philanthropy, and policy.

In the tradition of the New Directions series, each volume of the journal addresses a single, timely topic, although special issues covering a variety of topics are occasionally commissioned. We welcome submissions of both volume topics and individual articles. All articles should address the implications of theory for practice and research directions, and how these arenas can better inform one another. Articles may focus on any aspect of youth development; all theoretical and methodological orientations are welcome.

If you would like to be an *issue editor*, please submit an outline of no more than four pages that includes a brief description of your proposed topic and its significance along with a brief synopsis of individual articles (including tentative authors and a working title for each chapter).

If you would like to be an *author*, please submit first an abstract of no more than 1,500 words. Send this to the editorial manager.

For all prospective issue editors or authors:

- Please make sure to keep accessibility in mind, by illustrating theoretical ideas with specific examples and explaining technical terms in nontechnical language. A busy practitioner who may not have an extensive research background should be well served by our work.
- Please keep in mind that references should be limited to twenty-five to thirty. Authors should make use of case examples to illustrate their ideas, rather than citing exhaustive research references. You may want to recommend two or three key articles, books, or Web sites that are influential in the field, to be featured on a resource page. This can be used by readers who want to delve more deeply into a particular topic.
- All reference information should be listed as endnotes, rather than including author names in the body of the article or footnotes at the bottom of the page. The endnotes are in APA style.

Please visit http://www.pearweb.org for more information.

Gil G. Noam
Editor-in-Chief

NEW DIRECTIONS FOR YOUTH DEVELOPMENT

ORDER FORM SUBSCRIPTION AND SINGLE ISSUES

DISCOUNTED BACK ISSUES:

Use this form to receive 20% off all back issues of *New Directions for Youth Development.*
All single issues priced at **$23.20** (normally $29.00)

TITLE	ISSUE NO.	ISBN
_____	_____	_____
_____	_____	_____
_____	_____	_____

Call 888-378-2537 or see mailing instructions below. When calling, mention the promotional code JB9ND
to receive your discount. For a complete list of issues, please visit www.josseybass.com/go/ndyd

SUBSCRIPTIONS: (1 YEAR, 4 ISSUES)

☐ New Order ☐ Renewal

U.S.	☐ Individual: $85	☐ Institutional: $228
Canada/Mexico	☐ Individual: $85	☐ Institutional: $268
All Others	☐ Individual: $109	☐ Institutional: $302

Call 888-378-2537 or see mailing and pricing instructions below.
Online subscriptions are available at www.interscience.wiley.com

ORDER TOTALS:

Issue / Subscription Amount: $ _____

Shipping Amount: $ _____
(for single issues only – subscription prices include shipping)

Total Amount: $ _____

SHIPPING CHARGES:

First Item	$5.00
Each Add'l Item	$3.00

(No sales tax for U.S. subscriptions. Canadian residents, add GST for subscription orders. Individual rate subscriptions must be paid by personal check or credit card. Individual rate subscriptions may not be resold as library copies.)

BILLING & SHIPPING INFORMATION:

☐ **PAYMENT ENCLOSED:** *(U.S. check or money order only. All payments must be in U.S. dollars.)*

☐ **CREDIT CARD:** ☐ VISA ☐ MC ☐ AMEX

Card number _____ Exp. Date_____

Card Holder Name_____ Card Issue # _____

Signature _____ Day Phone_____

☐ **BILL ME:** *(U.S. institutional orders only. Purchase order required.)*

Purchase order # _____
Federal Tax ID 13559302 • GST 89102-8052

Name_____

Address_____

Phone_____ E-mail_____

Copy or detach page and send to: **John Wiley & Sons, PTSC, 5th Floor**
989 Market Street, San Francisco, CA 94103-1741

Order Form can also be faxed to: **888-481-2665**

PROMO JB9ND

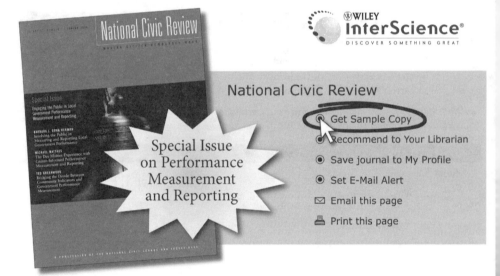